Building Pathology and Rehabilitation

Volume 17

This book series addresses the areas of building pathologies and rehabilitation of the constructed heritage, strategies, diagnostic and design methodologies, the appropriately of existing regulations for rehabilitation, energy efficiency, adaptive rehabilitation, rehabilitation technologies and analysis of case studies. The topics of Building Pathology and Rehabilitation include but are not limited to - hygrothermal behaviour - structural pathologies (e.g. stone, wood, mortar, concrete, etc…) - diagnostic techniques - costs of pathology - responsibilities, guarantees and insurance - analysis of case studies - construction code - rehabilitation technologies - architecture and rehabilitation project - materials and their suitability - building performance simulation and energy efficiency - durability and service life.

More information about this series at http://www.springer.com/series/10019

João M. P. Q. Delgado
Editor

Efficient and Suitable Construction

Springer

Editor
João M. P. Q. Delgado
CONSTRUCT-LFC, Department of Civil
Engineering, Faculty of Engineering
University of Porto
Porto, Portugal

ISSN 2194-9832 ISSN 2194-9840 (electronic)
Building Pathology and Rehabilitation
ISBN 978-3-030-62828-4 ISBN 978-3-030-62829-1 (eBook)
https://doi.org/10.1007/978-3-030-62829-1

This Springer imprint is published by the registered company Springer Nature Switzerland AG
The registered company address is: Gewerbestrasse 11, 6330 Cham, Switzerland

Preface

Since 50 years ago, the energy demand from buildings (residential and commercial) has grown by 1.8% per year, and it is predicted to grow from 2790 Mtoe, i.e. 116.8 EJ, in 2010 to over 4400 Mtoe by 2050, with most of this increase being from developing countries. Three-quarters of total energy consumption in the building sector is residential, where there is a great potential to improve energy efficiency. There is a presumption and need for these requirements to be applied not only to new buildings but also to the existing ones.

Ten years ago, this objective was viewed as unrealistic. Now, with emerging materials for thermal energy management, the PV panel reduction of costs and the development electrochemical storage energy (batteries) and simulation technologies, suddenly researchers and investors and consequently politicians begun to see this directive as possible, necessary and potentially interesting to invest.

The improvement of energy efficiency, comfort, hygrothermal behaviour and health in buildings is based on the application of building physics principles. The construction of industry efficiency is attained through process improvement, life cycle assessment, and smart management and maintenance.

The main purpose of this book, *Efficient and Suitable Construction*, is to provide a collection of recent research works related to hygrothermal building performance, acoustic and natural lighting performance in buildings, phase change material (PCM) and energy storage.

The book is divided into five chapters that intend to be a resume of the current state of knowledge for benefit of professional colleagues, scientists, students, practitioners, lecturers and other interested parties to network. At the same time, these topics will be going to the encounter of a variety of scientific and engineering disciplines, such as civil, mechanical and materials engineering.

Porto, Portugal João M. P. Q. Delgado

Contents

Phase Change Materials: From Fundamentals and Melting Process to Thermal Energy Storage System for Buildings Application

T. N. Porto, J. M. P. Q. Delgado, A. S. Guimarães, A. G. Barbosa de Lima, T. F. Andrade, H. L. F. Magalhães, G. Moreira, and B. B. Correia

Abstract This chapter is devoted to study the melting process of a phase change material into a triplex tube heat exchanger with finned copper tubes. The focus is on the heat transfer between a heating fluid (water) and phase change material to power a liquid-desiccant air conditioning system. Here, different topics related to fundamentals and applications of the phase change materials, and storage energy system with especial reference to a triplex tube heat exchanger are presented and discussed. Results of the temperature fields, liquid fraction, and velocity, as well as the energy history of the involved phases in the heating and melting processes are presented, compared with experimental data, and analyzed.

T. N. Porto · A. G. Barbosa de Lima (✉) · B. B. Correia
Department of Mechanical Engineering, Federal University of Campina Grande, Campina Grande 58429-900, Brazil
e-mail: antonio.gilson@ufcg.edu.br

T. N. Porto
e-mail: trnporto@gmail.com

B. B. Correia
e-mail: balbinacorreia@hotmail.com

J. M. P. Q. Delgado · A. S. Guimarães
Department of Civil Engineering, Faculty of Engineering, CONSTRUCT-LFC, University of Porto, 4200-465 Porto, Portugal
e-mail: jdelgado@fe.up.pt

A. S. Guimarães
e-mail: anasofia@fe.up.pt

H. L. F. Magalhães · G. Moreira
Department of Chemical Engineering, Federal University of Campina Grande, Campina Grande 58429-900, Brazil
e-mail: hortencia.luma@gmail.com

G. Moreira
e-mail: gicelia.moreira@eq.ufcg.edu.br

T. F. Andrade
Department of Petroleum Engineering, Federal University of Campina Grande, Campina Grande 58429-900, Brazil
e-mail: tonyherbert2000@gmail.com

J. M. P. Q. Delgado (ed.), *Efficient and Suitable Construction*, Building Pathology and Rehabilitation 17, https://doi.org/10.1007/978-3-030-62829-1_1

Keywords PCM · Thermal energy storage · Energy demand · CFD analysis

1 What Are Phase Change Materials?

1.1 Definition and Classification

Materials capable of absorbing or releasing large amounts of energy, at certain periods of time and under specific operating conditions are called Phase Change Materials (PCMs). These materials are capable of storing 5–14 times more energy per unit of volume than materials that store energy via sensitive heat, such as water, concrete or rocks, and present specific phase-changing temperatures that tend to remain constant during the transformation of matter (Sharma and Chen 2009).

Depending on the type of PCM, the energy storage process can be described by a transition between the solid–solid phases (changes in the crystalline structure of a material, which characterizes the storage or release of energy), solid–liquid, liquid–gas, or solid–gas. However, liquid–gas and solid–gas transformations are not applied to building materials due to their high volume and pressure variations during phase change processes. Solid–solid transformations are limited due to the difficulty of mixing with other building materials, such as cement and gypsum. Therefore, solid–liquid processing PCMs are most used in latent thermal storage systems (Su et al. 2015).

The operating cycle of a solid–liquid transition PCM is divided into two distinct periods, the heating phase (load) and the cooling phase (discharge). During the periods in which the system is receiving heat from a given thermal source, the material is heated and, reaching the melting temperature, it is characterized by a solid–liquid biphasic zone, which is exceeded when, in its entirety, the material reaches the liquid state. From then on, the PCM is heated via sensitive heat. When the power supply of the system is cut off, the process of energy discharge of the material begins. In this period, the PCM functions as a power source of the system, cooling until it reaches its melting temperature, were passing through the biphasic zone and, at the end of the solidification process, returning the initial temperature of the system (Antunes 2016).

The PCMs are subdivided into organic, inorganic and eutectic, and organic, subdivided into paraffinics and non-paraffinics (fatty acids, alcohols and glycols); inorganic, in turn, are separated into salt hydrates, organic compounds and metals. Eutectics, on the other hand, represent the mixtures that can occur among PCMS, which are organic or inorganic.

Organic materials are not corrosive, have low or no undercooling, are chemically and thermally stable, to the point of undergoing several cycles of fusion and solidification without being segregated, which represents a degradation of the latent heat of the material. On the other hand, these PCMs have low thermal conductivity and are flammable (Zalba et al. 2003; Sharma et al. 2009, 2015; Su et al. 2015).

Paraffin's belong to the family of saturated hydrocarbons (linear chain n-alkanes), which are represented by the chemical formula C_nH_{2n+2}, and the number of carbon atoms may vary between 20 and 40. Depending on the length of the alcanchain, the paraffin's can be of equal chain (n-paraffin) or odd chain (iso-paraffin). As the number of carbon atoms is increased, the melting temperature increases and a greater amount of latent heat is required to promote the phase transition of these materials, which occurs due to the high induced dipolar attraction, existing between the n-alkane chains. Paraffin's are capable of storing and releasing a large amount of heat and supporting a large number of phase change cycles; are excellent materials for energy storage, having specific heats ranging from 2.10 to 2.21 kJ/kg K, thermal conductivity between 0.132 and 0.346 W/mK; and latent melting heat between 147 and 266 kJ/kg (Himran et al. 1994).

Non-paraffinics organic materials such as esters, fatty acids, alcohols and glycols can also be used for energy storage via latent heat, especially fatty acids. However, they are flammable materials and should not be exposed to excessively high temperatures, flames or oxidizing agents. These materials have high latent melting heat and flammability; have low thermal conductivity and flash point, variation in toxicity level and instability at high temperatures (Su et al. 2015).

Inorganic phase-shifting materials are classified into salt hydrates, metals and inorganic compounds. These PCMs cover a wide range of operating temperatures and, due to their high density, fusion enthalpy, for a given volume, are higher compared to organic materials. Its main disadvantage is the reactivity with metals, and may have severe levels of corrosion being developed, in some combinations of metal with PCM.

Eutectic PCM's are composed of minimal fusion of two or more components, each of which congruently merge and solidify forming a mixture during solidification. Eutectic materials almost always merge and solidify without segregation, since they have an intimate mixture of crystals, leaving little opportunity for the components to separate. When merging, both components liquefied simultaneously, again with unlikely separation (Sharma et al. 2009). The main advantage of eutectics over other types of PCMs is that their melting points can be adjusted by combining different percentages of component PCMs. Eutectic also have high thermal conductivity and density, and do not experience super-cooling. However, the calorific capacity and specific heat are much lower than salt hydrates and paraffin's (Su et al., 2015).

According to Zalba et al. (2003), the storage of thermal energy, through phase-changing materials, is presented as a solution to essential issues, in the search for increased energy efficiency of thermodynamic systems, namely the delay between the availability of energy and its consumption in receiver systems (solar energy and cogeneration systems), the guarantee of energy in essential sectors, such as hospitals, inertia and thermal protection. In the first case, PCMs can be associated with solar energy absorption plants or power systems, which lose thermal energy to the environment in the form of heat. During the time periods when solar thermal energy is available, the phase change material is charged.

In the case of cogeneration systems, this would occur when the system was releasing heat to the PCM. At times when solar power is unavailable or below the desired amounts, the PCM would release power to the system. The same would happen when, in the cogeneration systems, the stored energy was directed to the increase of its power. In the second case, phase change materials would serve as thermal energy reserve sources for systems that need to have their temperature controlled; this occurs through bursting energy sources such as solar or electric. Through thermal interactions with these systems, the energy stored in the PCMs can be used to ensure the fulfilment of your energy demands.

1.2 PCM Applications

As previously mentioned, depending on the type of PCM, energy storage can occur due to the phenomenon of transition between phases of solid–solid, solid–gas, liquid–gas, and solid–liquid types.

The applicability of solid–liquid transition PCMs depends on several factors, such as: application-appropriate phase transition temperature, high latent heat per unit weight, high thermal conductivity and calorific capacity. In addition, it is desirable that the PCM has a favourable equilibrium phase for its application, high density, and minimal volume variation during the phase transition, and low steam pressure at operating temperature (Su et al. 2015).

Phase change materials can be used both to isolate and to control the temperature of certain environments, such as greenhouses or homes, which are subject to variations in thermal conditions, external to them. During sunny days, PCMs can be charged through solar collectors and this energy can be released into these environments in periods of low temperatures, such as at night. In addition, they can function as thermal insulators, with greater capacity than materials intended for this function, and that use sensitive heat to do so. The following are some applications of PCM materials.

(a) **Solar water heating**: In these systems, a PCM layer fills the bottom of a water heating reservoir. During the day, the water heat up via solar thermal energy and starts to transfer heat to the PCM. This, in turn, absorbs energy in the form of latent heat and fuses. At the end of the day, hot water is removed and replaced by cold water, which begins to receive energy from the heated PCM, which cools until it reaches its state of solidification (Sharma and Chen 2009).
(b) **Air heating by solar thermal energy** Waqas and Kumar (2013): used a system consisting of a solar collector, fans, which force air passage through the collector and energy storage units. During the day, the thermal energy obtained in the solar collector heats the air, which passes through the energy storage unit, bringing the PCM to its liquid state. This heat exchange causes the air to cool down in the environment. During the night, the PCM is cooled with cold air, characteristic of this period, reaching the state of solidification and causing the air to enter heated in the proposed environment.
(c) **Solar oven**: Solar ovens are able to operate overnight or on cloudy days. These equipments operate by storing thermal energy from the phase change of a PCM,

allocated in an oven cavity. To reach high temperatures, lenses are used that focus on the incident sun rays in the oven throughout the day (Sharma et al. 2000).

(d) **Solar greenhouse**: PCMs are also used for maintaining temperatures in greenhouses. Sharma et al. (2009) developed a greenhouse with energy storage units via PCM. The storage units absorb heat from the heated air, which enters the greenhouse during the day, through the fusion of the PCM. During the night, this energy is released inside the greenhouse. In addition, the PCM also functions as a thermal insulator, absorbing heat from hot air, which is released at the outlet of the greenhouse.

(e) **Building construction**: The building materials developed with PCMs are applied in order to control the temperatures in the environment of the residences throughout the day. Constructions that use solar heating with latent heat storage via PCMs show a reduction in the dimensions of the walls. This occurs due to the high thermal capacity and characteristic densities of these materials. Another aspect that can be mentioned in relation to buildings is the transfer of heat by radiation, which the walls composed of PCM start to release at night, to the residential environment, promoting the heating of air inside the residence (Kośny 2015). Due to its importance, several studies related to the application of PCM materials under construction have been reported in the literature (Kuznik et al. 2011; Zhou et al. 2012; Memon et al. 2014; Kenisarin and Mahkamov 2016; Wahid et al. 2017; Mousa et al. 2020 and Vaz Sá et al. 2020).

2 Storage Energy System: Triplex Tube Heat Exchanger

2.1 Fundamentals

Triplex tube heat exchangers (TTHX) have strong application in the pharmaceutical, food and beverage industries (García-Valladares 2004). These heat exchangers are constructed with three concentric tubes (internal, intermediate and external), constituting 3 distinct volumes in the exchanger, namely an external annular, between the external and intermediate tubes, an intermediate annular, between the intermediate and internal tubes and a cylinder, corresponding to the interior of the inner tube. Through these volumes circulate fluids of thermal exchange of cooling (Cold Heat Transfer Fluid-CHTF) or heating (Hot Heat Transfer Fluid-HHTF), which are responsible for loading or unloading the PCM. In order to intensify the thermal flow between the PCM and the HTF, the surfaces of the pipes in contact with the PCM are flipped, act since one of the limitations of the application of these materials is the low thermal conductivity.

The loading and unloading processes of the PCM occur in different periods or simultaneously: In the first situation, during the loading process, the HHTF enters a piping system, which allows it to heat the PCM, exchanging heat by convection, from the internal surfaces of the inner and outer tube of the intermediate tube, with the aid of fins, manufactured with high thermal conductivity material. This process causes the PCM to merge. At the end of the fusion process, the discharge process begins, injecting the CHTF into the heat exchanger, where the PCM is brought to solid state, releasing energy to Latent Heating Thermal Energy System (LHTES). For both processes occurring simultaneously, the cold heat transfer fluid (CHTF) is drained through the external annular; through the inner tube, drains a heat exchange fluid (Hot Heat Transfer Fluid-HHTF). The PCM, on the other hand, is confined to the intermediate annular that continuously stores thermal energy from the HHTF and releases this energy to the CHTF, through the solidification process on the one hand and the fusion process on the other.

2.2 *Heat Transfer and Storage Enhancement Methods*

2.2.1 Basic Concepts

The storage of thermal energy can occur by means: (a) heat from reversible chemical reactions, (b) sensitive heat, which allows liquid or solid media to increase their internal energy, raising their temperature and (c) latent heat, characteristic of the phase change of some materials that, when subjected to thermal exchange, modify the state of the matter that composes them, increasing its internal energy and maintaining its temperature, practically constant.

Sensitive heat storage is easily implemented by heating or cooling substances; however, the low thermal storage capacity and the high weight or volume required for this type of storage limit your applications. The storage of chemical energy has not been constantly applied in practice due to both technical issues necessary for its application and economic issues (Zhang et al. 2016). Therefore, storage in the form of latent heat stands out, due to the high heat transfer rates associated with relatively low volumes, which this type of storage is capable of promoting.

Over the past 36 years, the storage of thermal energy through the transformation of material phases has been an important topic, addressed in research, globally. Works, such as Abhat (1983) and Lane (1983), were pioneers in studying the ability of various materials to absorb and release large amounts of energy in certain periods of time and under specific operating conditions.

The steps in the development of Latent Heating Thermal Energy Storage built with PCM are research themes that develop in the area of material sciences and transport phenomena. For a given application, an LHTES must be constructed from two fundamental segments:

(a) The definition of the material that fits the operating conditions in which energy is intended to be stored. In this sense, studies have focused on the characterization of PCMs and other component materials, used in the construction of the thermal storage system (Abhat 1983; Lane 1983; Zalba et al. 2003; Sharma and Chen 2009; Costa 2014; Sharma et al. 2015; Zhang et al. 2016). In these studies, the thermophysical properties, the compatibility between PCMs and building materials and their behaviour are analysed throughout the loading and unloading cycles of the LHTES, taking into account questions about the life of the system components.

(b) Starting from the results that characterize the materials involved, there is the development of heat exchangers that use, with maximum efficiency, the properties of the phase change materials (García-Valladares 2004; Li et al. 2012; Liu et al. 2012; Al-Abidi et al. 2013; Karthikeyan et al. 2014; Tristão 2014; Li and Wu 2015; Esapour et al. 2016; Xu et al. 2016; Bahrani et al. 2017). In this context, the research is focused on the analysis of questions about the modes of heat transfer (conduction, convection and radiation) that are involved in the storage system, in the determination of research parameters that describe the thermal and fluid dynamic aspects, such as Reynolds (Re) and Nusselt (Nu) numbers, simulations, the development of empirical models (from experiments), and prototypes and pilot plants, which are tested and have their costs evaluated.

For the energy storage process to occur efficiently, the phase-changing material must be designed to absorb or release as much energy, within the time interval, in which it is being charged or discharged. Thus, except for metals, the low thermal conductivity of PCMs is characterized as a limiting factor for their application in thermal storage systems. In this sense, several studies have been carried out with the aim of developing techniques to increase/intensify the flow of heat to or from PCMs during the processes of energy storage and release (Hasnain 1998; Sharma et al. 2000; Peng et al. 2004; Liu et al. 2016; Ibrahim et al. 2017; Lin et al. 2018).

2.2.2 LHTES Enhancement Techniques

LHTES improvement techniques are subdivided between those designed to increase heat transfer in these systems and those designed to increase the thermal conductivity of PCMs.

Heat Transfer Intensification Techniques

(a) Extended surfaces

The use of extended surfaces is made in order to increase the interaction area (heat exchange) between the PCMs and the source of loading or unloading, associated with it. Thin structures extend from the walls that confine the phase-changing material, so as to increase the thermal exchange fronts, intensifying the heat transfer process in the storage systems. Several studies have been developed with the objective of determining the influence of the shape and types of fin materials used to intensify the fusion-solidification processes of PCMs (Al-Abidi et al. 2013; Jmal and Baccar 2015; Pizzolato et al. 2017; Abdulateef et al. 2018). The selection of the fin material must meet the needs of thermal conductivity, density, cost and corrosion potential, existing in the LHTES project (Liu et al. 2012).

Problems involving extended surfaces in LHTES are subdivided into two areas: Those that use a thermal exchange fluid (HTF) to heat or cool PCMs, used in solar water or air heaters, and those that do not use HTF, and PCMs function as heat sink reservoirs, used to cool heated systems such as electronic appliances. In these cases, the pin fins are located within the structure of the PCM that is put in contact with heated surfaces of electronic equipment. In cases where HTF is used, the fins are on the side of the fluid with the lowest thermal conductivity, which, in most cases, is the PCM, as in the TTHX type heat exchanger described in the previously described section.

(b) Heat pipes

Heat pipes (HP) are devices that use, in their structure, the condensation or evaporation of secondary HTFs, to intensify the thermal flow between the PCM and the main HTF (Nithyanandam and Pitchumani 2013). LHTES modules of the hull and tube type, for example, use heat pipes, as thermal flow intensifiers. The PCM may be filling the hull volume and the main HTF seeping inside the heat exchanger tubes, or the PCM may be filling the volume of the tubes and the main HTF, seeping into the hull volume. Heat tubes are distributed along the LHTES connecting the volumes filled by PCM and the main HTF. At the ends of the HP, condensation and evaporation of the working fluid (secondary HTF) occur, causing it to conduct thermal energy more intensely and accelerate the phase change process in the PCMs. If a given end is located in the PCM section and it is in the charging process, this will be the end where secondary HTF condensation will occur. On the other hand, at the end that is in the section referring to the main HTF, evaporation is occurring. The opposite will occur if the PCM is in the process of downloading.

Shabgard et al. (2019), when numerically studying an LHTES applied in solar thermal power plants, using both loading and unloading modes, reported a significant increase in energy storage and release rates due to the presence of HPs. Using a single heat tube, the researchers found that of the total energy stored in the PCM during the charging process, 13% was in the vicinity of HP, showing the relevant ability of this method, to accelerate the process of thermal exchange.

(c) Multiple PCMs

In this method of thermal exchange intensification, the same HTF, along its route in LHTES, exchanges heat with various phase change materials, which have different melting points. That is, even if the HTF loses thermal energy, heating the materials with the highest melting point, arranged at the beginning of the trajectory, it still has the ability to raise the temperature of the materials with the lowest melting point, which are sequentially allocated in the remainder of this trajectory, up to the phase transition temperature, promoting a greater efficiency of the thermal exchange process (Ibrahim et al. 2017).

Increased Thermal Conductivity

(a) Composition with high thermal conductivity materials

The addition of materials of high thermal conductivity consists in the manufacture of a composite material consisting of PCMs and metallic materials, carbon-based materials and other materials such as boron nitride and silica (Lin et al. 2018). These can be applied in the form of spherical or irregularly shaped particles, randomly dispersed in the matrix or using porous structures (Foams), which when immersed in PCMs (liquid or pasty), have their pores filled by PCM. From this, an interconnected network of phase change materials begins to interact thermally, from the structure of high conductivity materials. Oya et al. (2012) developed a composite material of nickel (Ni) porous metal foam (MF) impregnated with Erythritol ($C_4H_{10}O_4$) as a phase change material. Nickel has porosity of 84–86% and an average pore size of 500 μm, and this is filled under vacuum condition with a pressure of 5.0×10^{-4} Pa and a temperature of 150 °C.

(b) Encapsulation

In addition to low thermal conductivity, other issues such as high volumetric variations during phase change, leaks and flammability of organic PCMs, and corrosion, decomposition and sub-cooling, characteristic of inorganic PCMs, are issues that prevent their application. In this sense, the encapsulation techniques of these materials have been developed. Through physical–chemical methods it is possible to package fractions of phase-changing materials in capsules of materials of higher thermal conductivity, usually polymers or metals, increasing the heat transfer area and reducing the reactivity of PCMs in relation to the external environment (Zhao and Zhang 2011).

Encapsulation techniques can be subdivided according to the capsules size into macro or microencapsulation. Macroencapsulation consists of allocating the PCM in a macroscopic containment, containing from a few millilitres to several litters of material. These containers are often made of metal or plastic. This technique is common due to the high availability of these containers, which are already manufactured for other applications. In this way, macro-encapsulation is mainly done to keep the PCM in the liquid state, avoid changes in its composition due to contact with the environment and, if the container is resistant, the encapsulation can also add mechanical stability to LHTES (Cabeza and Mehling 2008).

Microencapsulation occurs when capsules with dimensions between 1 and 1000 μm in diameter are used to coat solid or liquid particles of PCMs. Physical processes used for microencapsulation are drying processes by atomization, centrifugal and fluidized bed, or coating processes. The chemical processes used are in-situ encapsulations such as complex coacervation of gelatin, interfacial polycondensation, to obtain a polyamide or polyurethane coating and others. These processes are capable of producing materials with high water tightness and higher heat transfer due to the large surface area/volume ratio of the capsules (Cabeza and Mehling 2008).

3 Application: Use of TTHX in Building Construction

3.1 Physical and Computational Domains

The proposed physical study was the analysis of the PCM (RT82) fusion process, which exchanges heat with the water (the HTF) through a TTHX made of copper tubes with eight longitudinal fins with a 1 mm thickness and 42 mm length installed on the tubular surfaces, and was directly in contact with the PCM. The study is adapted from the work of Porto et al. (2020) published under an open access Creative Common CC BY license. As illustrated in Fig. 1, the TTHX was described using five physical domains, namely, the fluid volume (water) circulating in the pipes, the solid volumes (internal, intermediate, and external copper tubes), and the PCM (the volume between the intermediate tube inner surface and the inner tube outer surface). In the energy storage process, water entered the heat exchanger with a pre-established temperature T_i and mass flow rate $\dot{m}$ through a copper tube with a 50.8 mm diameter and a thickness of 1.2 mm. The mass flow rate is divided by a branch in the inlet pipe, which was 32 mm in diameter and 1.2 mm thick. Thus, the HTF transferred energy to the PCM through the intermediate tube's outer surface and the inner tube's internal surface. Then, the heat was conducted through the finned pipes to the PCM. As the HTF released energy to the system, its temperature at the outlet T_s was reduced. The TTHX external surfaces were considered to be isolated such that the heat transfer with the external environment was neglected.

In Figs. 2, 3, 4, 5, 6 and 7, the frontal and lateral views of the TTHX, frontal and isometric views of the fluid domain, of the internal and intermediate tubes, and the PCM with its dimensions are presented, respectively.

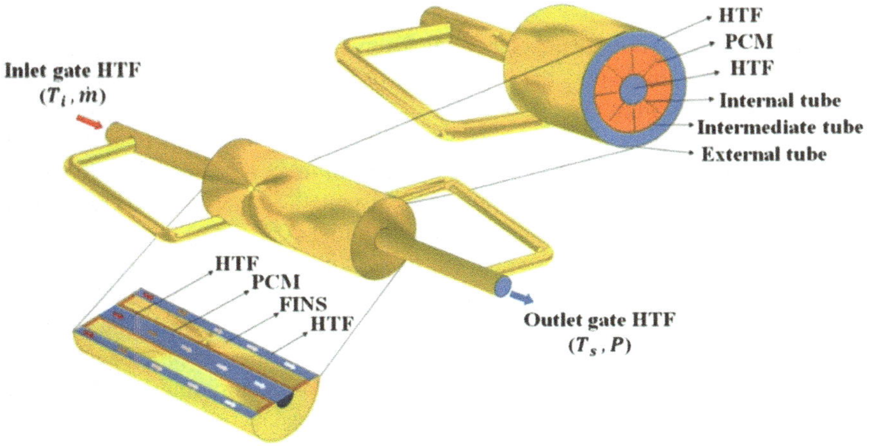

Fig. 1 Physical problem. HTF: Heat transfer fluid, PCM: Phase change material, TTHX: Triplex tube heat exchanger

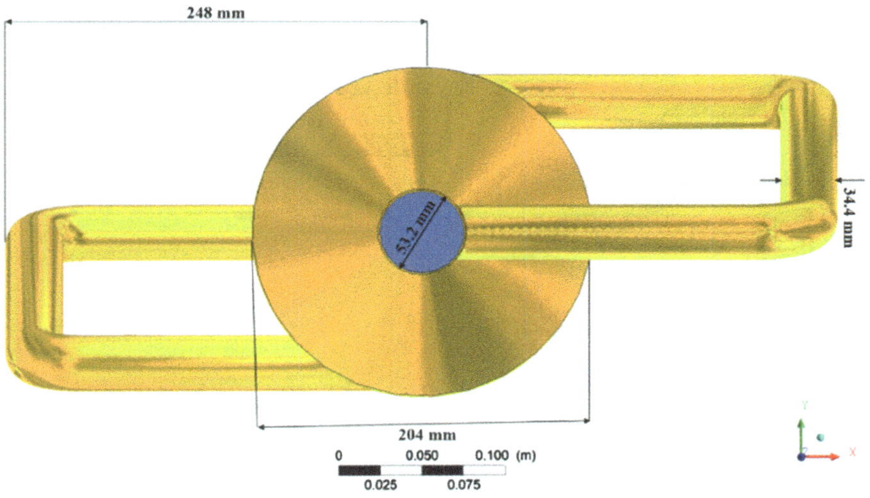

Fig. 2 TTHX front view

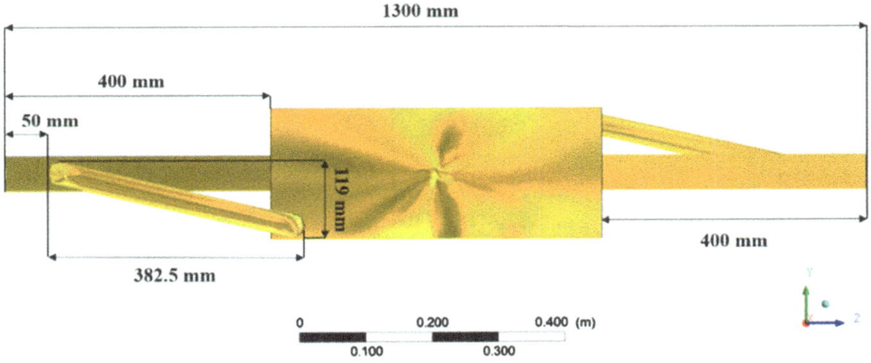

Fig. 3 TTHX side view

To simplify, the domain referring to the external tube was not considered in the simulations due to its small thickness and the insulation considered on the TTHX external surface. In this way, only the domains referring to the HTF, PCM, and internal and intermediate pipes were considered.

As can be seen in Fig. 5, there was a variation of 16 mm between the tube length and the fins' length. Adding the thickness (2 mm) of the neglected external tube, which lined the complete heat exchanger, this difference grew to 20 mm, with 10 mm at each end of the heat exchanger. Another simplification used was to consider only the 480 mm of the TTHX length, corresponding to the fin length.

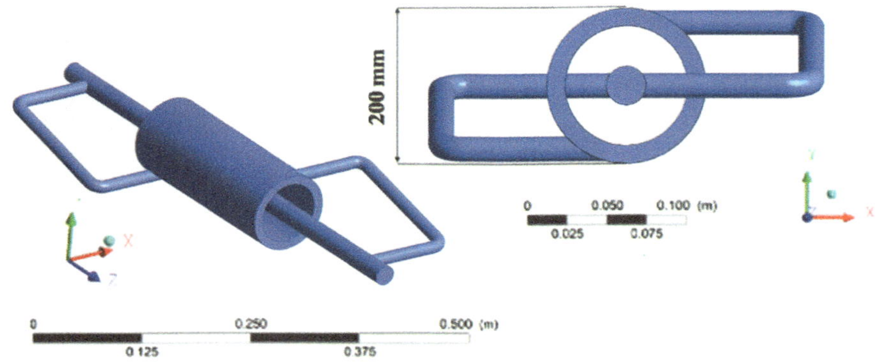

Fig. 4 Fluid domain front and isometric view (HTF)

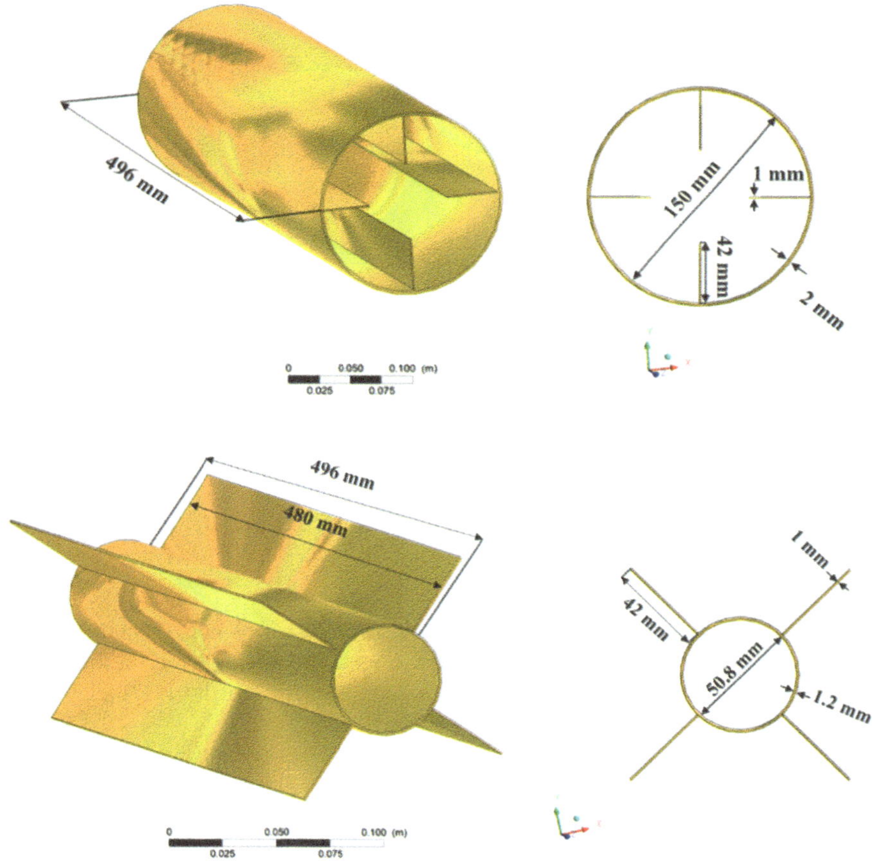

Fig. 5 Front and isometric view of the geometric domains referring to the intermediate and internal tubes

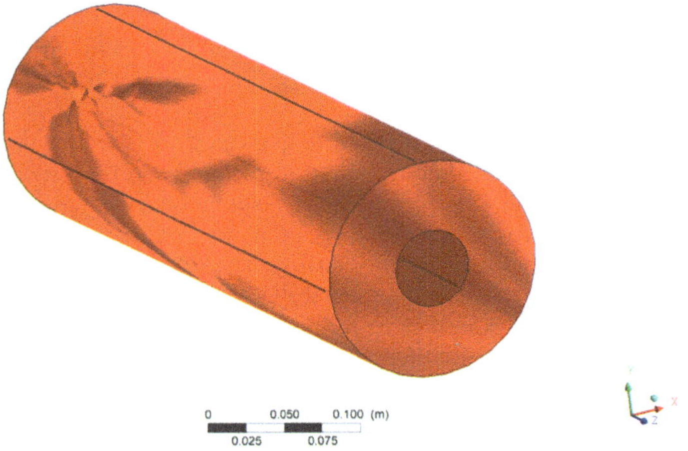

Fig. 6 Phase change material domain

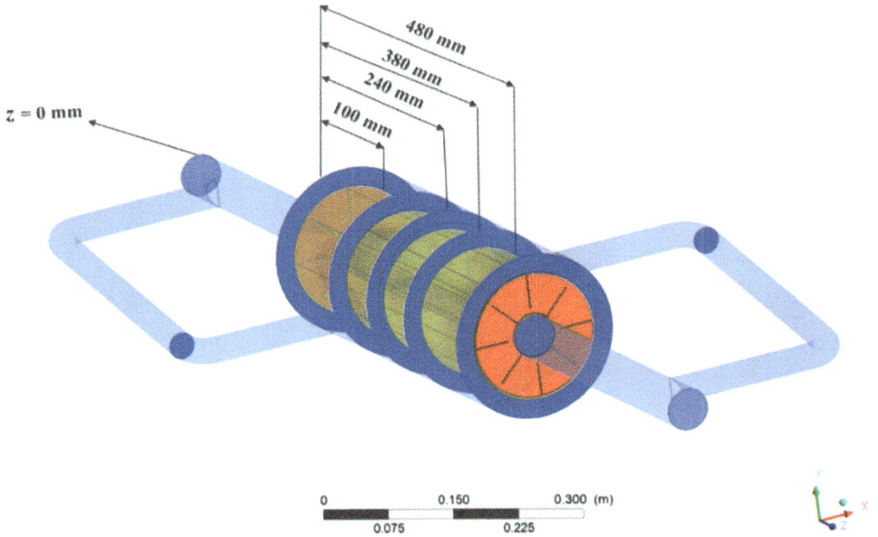

Fig. 7 Studied domains and analyzed planes

Thus, the simplified physical problem took the form showed in Fig. 7, where the five planes along the length of the exchanger, which were treated in the results presented in this work and were located at $z = 0$, 100, 240, 380, and 480 mm, are presented. These simplifications allowed for the production of meshes with considerable quality. The presented geometry was performed in ANSYS Design Modeler® software (Canonsburg, Pennsylvania, USA).

The heat exchange area (finned pipe surfaces in contact with the PCM) was 0.308832 m^2. The volume occupied by the PCM was 0.00725707 m^3. Converting these values to mass, depending on the PCM density, there was 6.89 kg of material stored in the equipment.

In this work, the finite volume discretization method was used. Hybrid unstructured meshes (tetra and hexahedral elements) unstructured were developed within the scope of the Ansys Meshing® software. This type of mesh made it possible to associate the molding capacity with the complex geometries of the tetrahedral elements allocated in the HTF domain, where good quality the results were related to the hexahedral elements in the finned structures and the PCM domain, which maintained the mesh quality close to the wall regions for that geometry. Figures 8, 9, 10, 11, 12 and 13 illustrate the meshes used for the domains treated in the physical problem. Figures 8 and 9 show the isometric and front view of the mesh used for the heat exchanger with all domains assembled.

Figure 10 presents the details of the mesh used for HTF, which had tetrahedral elements inside the volume and layers of hexahedral elements on the walls due to the hydrodynamic and thermal boundary layers developed in these regions.

Figures 11, 12 and 13 show the domain of the internal and intermediate finned pipes, as well as the PCM. Through the procedure used, it was possible to maintain element uniformity in the pipes, while in the PCM, the elements had an irregular shape that could track the PCM geometry.

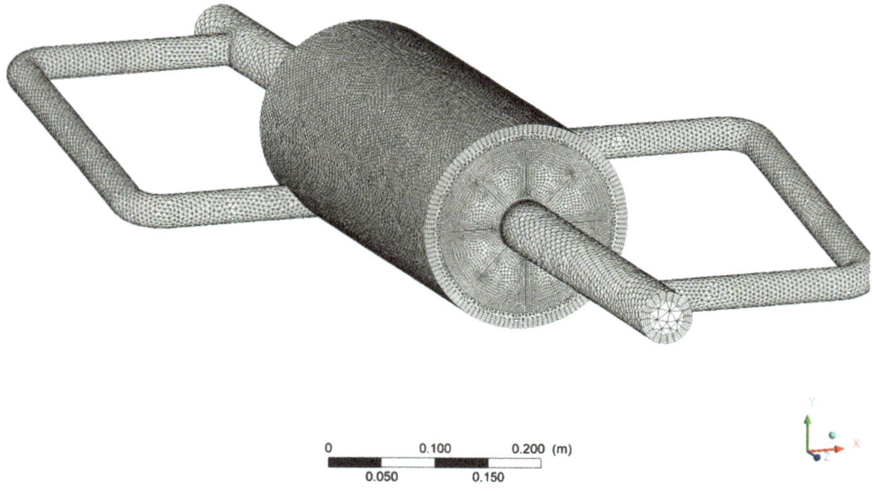

Fig. 8 Isometric view of the mesh used for the TTHX

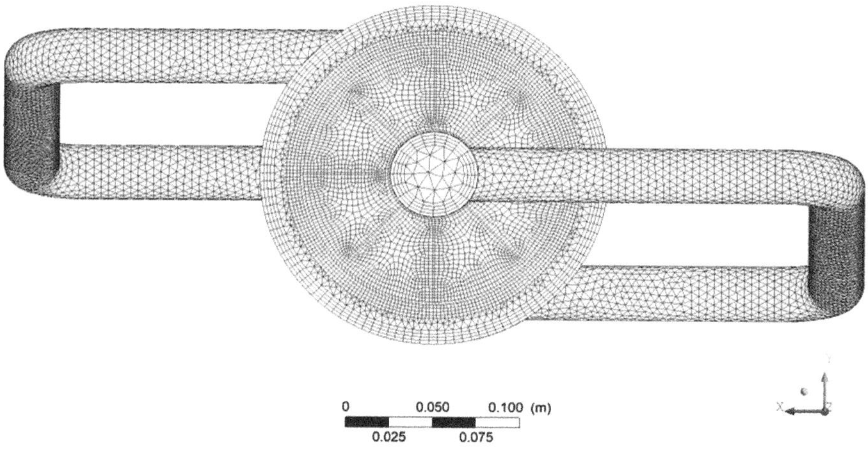

Fig. 9 Front view of the mesh used for the TTHX

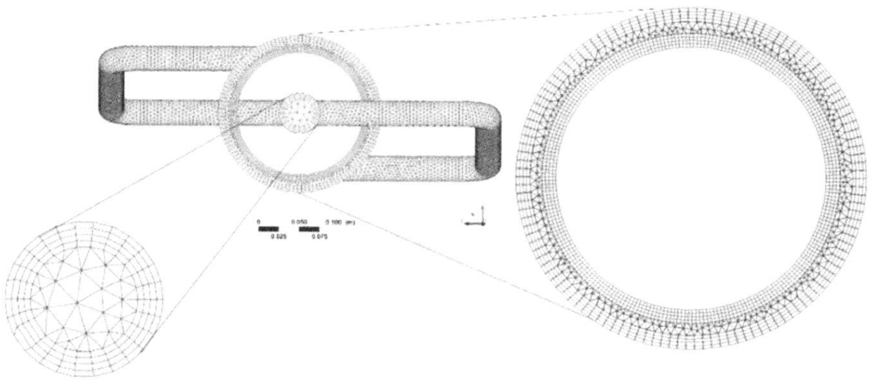

Fig. 10 Details of the mesh used for the HTF

3.2 Mathematical Modeling

The conservation laws that described the operation of the TTHX are presented in Eqs. (1)–(10). When modeling, the analysis must include the fluid dynamics, and thermal and phase transition processes, which occur simultaneously during PCM loading.

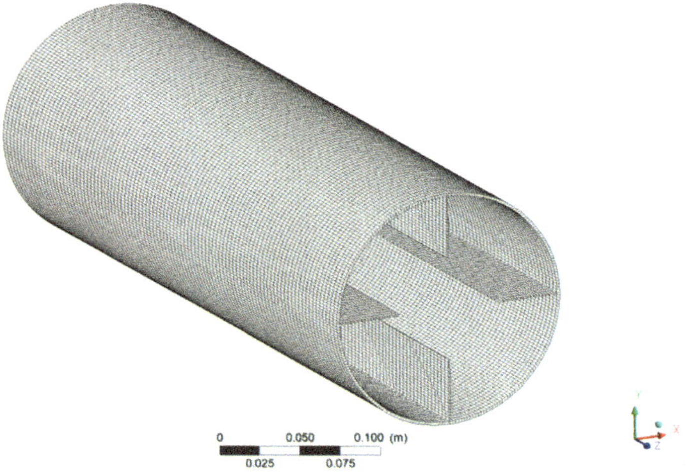

Fig. 11 Mesh used for the intermediate tube domain

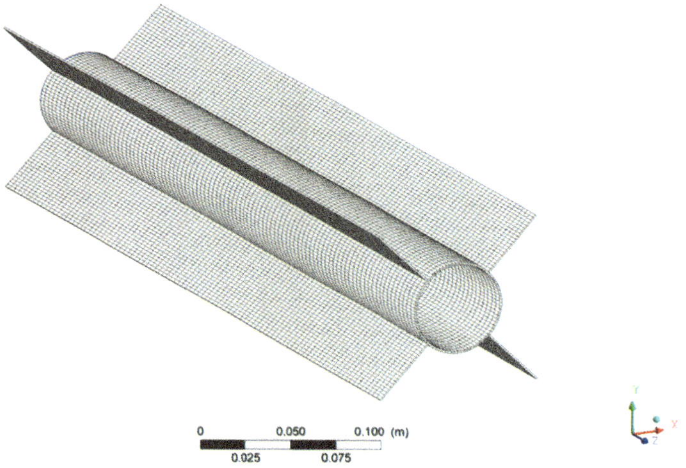

Fig. 12 Mesh used for the inner tube domain

(a) Mass conservation

$$\frac{\partial(\rho)}{\partial t} + \frac{\partial}{\partial X_i}(\rho u_i) = 0 \tag{1}$$

where ρ is the density of the fluid, t is the time variable, X is the position vector, u is the velocity vector, and the sub-indications i and j represent the components (x, y,

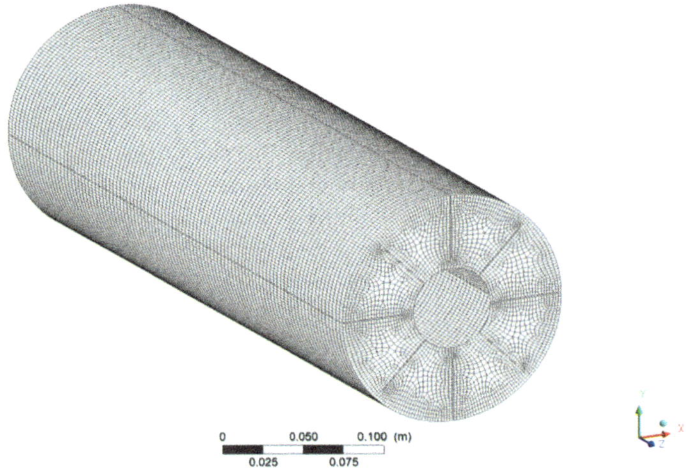

Fig. 13 Mesh used for the PCM domain

and z) of the coordinate axes such that i or $j = 1$ represents the x-direction, i or $j =$ 2 represents the y-direction, and i or $j = 3$ represents the z-direction.

(b) Linear momentum conservation

$$\frac{\partial(\rho u_i)}{\partial t} + \frac{\partial}{\partial X_j}(\rho u_i u_j) = -\frac{\partial P}{\partial X_i} + \frac{\partial}{\partial X_j}\left[\mu\left(\frac{\partial u_i}{\partial X_j} + \frac{\partial u_j}{\partial X_i} - \frac{2}{3}\delta_{ij}\frac{\partial u_i}{\partial X_i}\right)\right] + \frac{\partial}{\partial X_j}\left(-\rho\overline{u_i' u_j'}\right) \quad (2)$$

where μ is the dynamic viscosity, P is the pressure, and the term and $(\rho\overline{u_i' u_j'}$. represents the Reynolds stresses, which were derived from the turbulent flow.

(c) Turbulence through the shear–stress transport k-ω SST model

The stress transport model (shear–stress transport k-ω shear stress transport (SST) applied in the Ansys FLUENT® software was developed by Menter (1994) to combine the robustness and accuracy of the results close to the wall using the standard k-ω model (Wilcox 1998) with the accuracy and simplicity of the k-ε model (Launder and Spalding, 1972) for the regions far from the wall. To this end, coupling functions are used to activate the k-ω and k-ε models in the cells near and far from the walls, respectively, of the computational domain. The variable k represents the turbulent kinetic energy and the variable ω represents the dissipation rate of this energy. Equations (3) and (4) describe the transport of these variables.

$$frac\partial(\rho k)\partial t + \frac{\partial(\rho k u_i)}{\partial X_i} = \frac{\partial}{\partial x}\left(\Gamma_k \frac{\partial k}{\partial X_j}\right) + G_k - Y_k + W_k, \quad (3)$$

$$frac\partial(\rho\omega)\partial t + \frac{\partial(\rho\omega u_i)}{\partial X_i} = \frac{\partial}{\partial X}\left(\Gamma_\omega \frac{\partial \omega}{\partial X_j}\right) + G_\omega - Y_\omega + D_\omega + W_\omega, \quad (4)$$

where G_k and G_ω represent the generation of k and ω, Γ_k and Γ_ω are the effective diffusivity, Y_k and Y_ω represent the dissipation of k and ω, the term D_ω represents the cross-diffusion, and W_k and W_ω represent the source terms of the referred equations. Equations (1)–(4) are applied in domains that present flow: the HTF that is injected into the heat exchanger flows in a turbulent, laminar, or transition flow regime depending on the analyzed region, and the PCM presents a low-speed flow during the phase change. In this way, the k-ω SST turbulence model was used with corrections for low Reynolds (Re) numbers, which enabled the flow analysis in all referred regimes.

(d) Energy conservation

$$\frac{\partial}{\partial t}(\rho E) + \frac{\partial}{\partial X_i}[u_i(\rho E + P)] = \frac{\partial}{\partial X_j}\left[\gamma_{eff}\frac{\partial T}{\partial X_j} + u_i\left(\tau_{ij}\right)_{eff}\right] + W_h, \quad (5)$$

where T is the temperature, W_h is the energy source term, E is the total energy (Eq. (6)), and γ_{eff} is the effective thermal conductivity, described in Eq. (8):

$$E = h - \frac{P}{\rho} + \frac{|u_i|^2}{2}, \quad (6)$$

where h is the sensible enthalpy, defined as:

$$h = h_{ref} + \int_{T_{ref}}^{T} c_p dT, \quad (7)$$

where h_{ref} is the reference enthalpy, T_{ref} is the reference temperature, and c_p is the specific heat of the fluid under constant pressure. In the FLUENT® standard scheme, h_{ref} and T_{ref} are 0 J/kg and 15 °C, respectively.

$$\gamma_{eff} = \gamma + \gamma_t, \quad (8)$$

where γ_t is the turbulent thermal conductivity, described in Eq. (9):

$$\gamma_t = \frac{c_p \mu_t}{\sigma}, \quad (9)$$

where μ_t is the turbulent viscosity and σ is the Prandtl number.

The term $\left(\tau_{ij}\right)_{\text{eff}}$ is the effective stress tensor given by:

$$\left(\tau_{ij}\right)_{eff} = \mu_{eff}\left(\frac{\partial u_j}{\partial X_i} + \frac{\partial u_i}{\partial X_j}\right) - \frac{2}{3}\mu_{eff}\frac{\partial u_k}{\partial X_k}\delta_{ij}, \quad (10)$$

where

$$\mu_{eff} = \mu + \mu_t. \tag{11}$$

Equation (5) is applied to all domains to describe the heat transfer process.

(e) Phase change model

To solve the PCM melting transient problem the enthalpy–porosity model developed by Voller and Prakash (1987) was chosen. In this technique, the solid–liquid interface is not explicitly tracked, where the calculated liquid fraction is the one that indicates its position throughout the phase change processes. The heat transfer in the PCM domain occurs due to the thermal diffusion mechanism and natural convection due to density variations, which is a function of the temperature and phase change levels. Equations (12)–(16) were used for this stage of the process:

$$\frac{\partial}{\partial t}(\rho H) + \frac{\partial}{\partial X_i}(u_i \rho H) = \frac{\partial}{\partial X_j}\left[\gamma \frac{\partial T}{\partial X_j}\right] + S_h, \tag{12}$$

where

$$H = h + \Delta H, \tag{13}$$

and

$$\Delta H = \psi L, \tag{14}$$

where

$$\psi = \begin{cases} 0 \;\; if \;\; T < T_{sol}, \\ 1 \;\; if \;\; T > T_{liq}, \\ \frac{T - T_{sol}}{T_{liq} - T_{sol}} \;\; if \;\; T_{sol} < T < T_{liq}, \end{cases} \tag{15}$$

where ψ is the liquid fraction; H is the total enthalpy of the material, which is computed as the sum of the sensitive enthalpies h and the latent enthalpy variation of the material ΔH; L is the latent heat of fusion; T_{sol} is the solidification temperature; and T_{liq} is the temperature at which all the material is in a liquid state.

The PCM density (ρ) was calculated using Eq. (16). This equation couples the Boussinesq natural convection model (Ansys 2015) for $T \geq T_l$ and the mixture model for $T_{liq} \geq T \geq T_{sol}$. If the material is below T_{sol} the density (ρ_{sol}) remains constant.

$$\rho = \begin{cases} \rho_{liq}\left[1 - \eta\left(T - T_{liq}\right)\right] \;\; if \;\; T \geq T_{liq}, \\ \rho_{liq}\psi + (1 - \psi)\rho_{sol} \;\; if \;\; T_{liq} \geq T \geq T_{sol}, \\ \rho_{sol} \;\; if \;\; T_{sol} \geq T, \end{cases} \tag{16}$$

where ρ_{sol} and ρ_{liq} are the densities of the material in the solid and liquid states, respectively, and η is the thermal expansion coefficient. As the PCM is heated, it shows variations in its density, which primarily occur in the regions closer to the pipe walls. These density variations, when subjected to the gravitational force, are converted into buoyant forces, which move the PCM inside the heat exchanger at a low velocity.

3.3 Boundary Conditions

The prescribed mass flow rate was defined at the TTHX inlet. The mass flow rate specification allowed for the total pressure to vary in response to the numerical solution. In this boundary condition, the absolute reference system, the direction of flow normal to the inlet surface, the turbulence intensity I (Eq. (17)) of 5%, and the turbulent viscosity ratio of $R_\mu = 10$ (Eq. (18)) were previously established. The HTF at 90 °C was injected from the inlet gate with a volumetric flow rate of 8.3 L/min.

$$I \equiv \frac{u'}{\overline{u}} \tag{17}$$

$$R_\mu = \frac{\mu_t}{\mu} \tag{18}$$

The outflow boundary conditions were established at the TTHX outlet gate. In this model, the necessary information (velocities, pressure, and temperature) were extrapolated from the conditions within the physical domain. Two main features were activated in the solver as soon as the outflow boundary condition was enabled: (1) null diffusive flow, which means that the outlet conditions were extrapolated from the internal domain and that these conditions had no influence on the upstream flow; and (2) for all flow variables, a general mass balance correction was made. The wall conditions were used to connect the fluid and solid regions and to incorporate the insulation on the HTF external surfaces. Non-slip boundary conditions were applied to the pipe walls and the fluid surfaces of the domains, which were taken as stationary surfaces with a negligible roughness.

Regarding the heat transfer, two distinct conditions were used: zero flux for the external surfaces of the HTF domain and the coupled condition for the solid–liquid interfaces existing between the external surface of the intermediate tube and the internal surface of the inner tube with HTF, as well as between the inner surface of the intermediate tube and the outer surface of the inner tube with the PCM. Under these conditions, the heat flux was calculated at the interfaces depending on the temperatures of the neighboring cells of both domains. Table 1 describes the border regions of the physical problem and its boundary conditions.

Table 1 Boundary conditions used in the FLUENT® software

Regions	Boundary condition
TTHX inlet gate	Mass flow
TTHX outlet gate	Outflow
HTF external surface	Isolated wall
Internal surface of the inner tube–HTF	Coupled wall
Internal surface of the intermediate tube–PCM	Coupled wall
External surface of the internal tube–PCM	Coupled wall
External surface of the intermediate tube–HTF	Coupled wall

3.4 Numerical Procedures

Different numerical treatments were used to solve the governing equations: (a) for the pressure–velocity coupling, the coupled method was used; (b) for the spatial discretization, the least-squares cell-based method (Anderson and Bonhaus 1994) was used to determine the gradient $\nabla\Phi$; (c) the quadratic upwind implicit differencing convective kinematics (QUICK) method (Leonard and Mokhtari, 1990) was used for the discretization of linear momentum equations and turbulent kinetic energy generation; (d) the first- and second-order upwind methods (Barth and Jespersen, 1989) were used to discretize the turbulent kinetic energy and thermal energy dissipation equations, respectively; (e) the PRESTO! method (ANSYS 2015) was used for pressure discretization; and (f) for the temporal discretization, the implicit first-order method was used in the simulations.

Given the long simulation time required to simulate the PCM transient fusion process, the cases were initialized in a steady state, without the energy and phase change models being enabled. This procedure allowed for the pressure and velocity fields in the flow in the HTF to be established. As the steady-state process reached convergence, the fluid dynamic results were used as an initial condition for the transient simulation with the energy and phase change models enabled.

The relaxation factors used are described in Table 2. The convergence criteria used in the simulations was 2×10^{-4} for velocities in the x- and y-directions, 10^{-4} for velocities in the z-direction, mass conservation, balance of turbulent generation energy, and balance of turbulent dissipation energy, and 10^{-5} for energy conservation.

3.5 Thermo-Physical Properties of the Materials

The physical properties of the materials used in the simulations are described in Table 3, which were taken from Al-Abidi et al. (2013).

Table 2 Relaxation factors applied to the simulations

Parameter	Value (−)
Density	1
Buoyancy Force	1
Turbulent kinetic energy	0.8
Turbulence dissipation rate	0.8
Turbulent viscosity	1
Liquid fraction	0.9
Energy	1
Momentum	0.75
Pressure	0.75
Courant number	200

Table 3 Thermo-physical data used in the simulation

Material	μ (Pa s)	ρ (kg/m^3)	L (J/kg)	c_p (J/kg K)	λ(W/m K)	T_{sol} (K)	T_{liq} (K)
Water	0.001003	998.2	–	4182	0.6	–	–
PCM	0.03499	950 (sol) 870 (liq)	201,643.8	2000	0.2	343.27	355.32
Copper	–	8978	–	381	387.6	–	–

sol solid phase; *liq* liquid phase

It was considered that in the initial condition, the entire domain regarding the heat exchange fluid was completely filled with HTF at 90 °C, and that the PCM and the pipes were at a temperature of 27 °C. The time step used in the transient simulation was 0.5 s.

3.6 Results Analysis

3.6.1 Mesh and Time Step Refinements and Validation

To achieve coherent results using computational fluid dynamics (CFD) tools, it is necessary to carry out a mesh and time step refinement studies, which compare the results, obtained using more refined meshes with results from other less refined ones such that convergence becomes independent of the number of elements in the mesh convergence. The results were analyzed for two meshes and three time steps. Considering the analysis of the obtained results, a mesh with 1,155,528 elements and 0.5 s time step, were chosen.

To validate the results presented in this work, they were compared with the experimental results presented by Al-Abidi et al. (2014), which used a TTHX with the same dimensions shown in Fig. 1 and using paraffin RT82 as the PCM. From this comparison, a good concordance between the predicted and experimental results was obtained. Details about these procedures (refinement studies and validation) can be found in the literature (Porto et al. 2020).

3.6.2 Thermo-Fluid Dynamic Analysis

HTF Velocity

Figure 14 illustrates the velocity distributions for the heat exchange fluid (HTF). The isometric view of this figure was evaluated over a global range using velocities between 0 and 0.0097 m/s. In the reference planes taken along the length of the exchanger's external annular region at $z = 100$, 240, and 380 mm, local velocity variations were evaluated, which are presented as vectors that illustrate the flow directions.

Due to differences in the values of the local velocities relative to the external annular and the internal tube, greater resistance to the fluid directed toward the external region was observed. This was due to the pipe's geometric configuration. The tubular T-joint, existing at $z = -350$ mm, branching toward the main line (50.8 mm in diameter) in the perpendicular direction with a 32 mm diameter pipe, affected the flow of the fluid to the external region of the TTHX. It was observed that the maximum velocity at the external annular was 0.007 m/s, which was about 13 times lower than the maximum value found in the region comprised by the inner tube (0.097 m/s).

Comparing the graphs of the local velocities in Fig. 14, it can be seen that the levels of this variable decreased since the planes were presented in the positive z-direction. This was due to the distance of these planes relative to the fluid injection point and the conversion of the mechanical effect due to the injection pressure into a swirling movement, causing the HTF at the outlet of the external annular to present low pressure levels. In the plane at $z = 100$ mm, the vectors indicated a counterclockwise flow. In the plane at $z = 240$ mm, the vectors indicated a clockwise flow in the upper region of the annular and a counterclockwise flow in the lower region. Both in the 240 mm and 380 mm planes, regions with vectors pointing in various directions were observed, indicating a kind of chaotic movement (swirling), which occurred due to the proximity to the exit section.

Figure 15 shows the streamlines that described the trajectory of the HTF. It should be noted that the tangential positioning of tubes relative to the external annular promoted a swirling flow from the inlet to the outlet of this system. The pressure force that introduced the fluid in this region was converted into centrifugal forces due to the curved walls of the annular space.

It can be seen that the fluid injection at the annular lower region and fluid exit at the upper region caused the HTF to be displaced clockwise and counterclockwise from half the length toward the exit, which caused the vector directions described in Fig. 14. It was observed in the analysis of the heat transfer that this flow behavior generated asymmetric variations in the PCM temperature and liquid fraction fields.

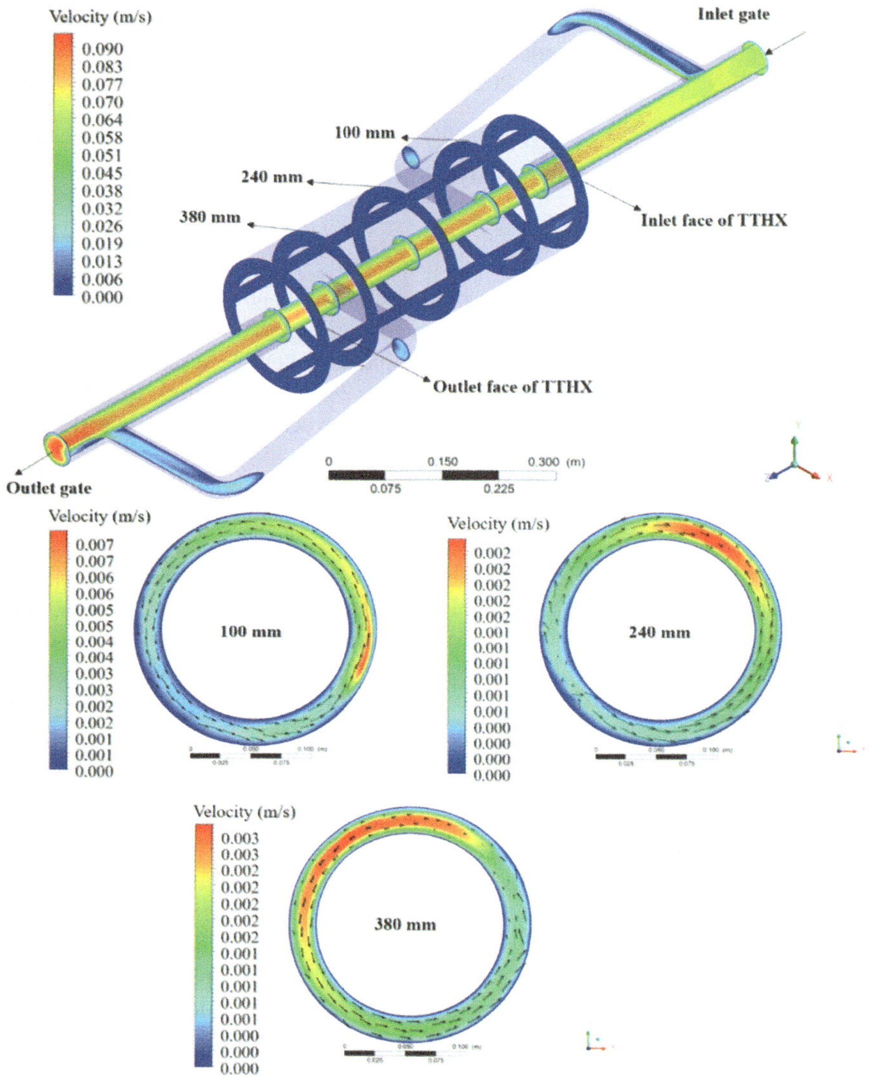

Fig. 14 HTF velocity field

Temperature Distribution

When considered together, Figs. 16, 17, 18 and 19 show the temperature distributions in all domains (heating fluid, the pipes, and the PCM) for four specific moments of the melting process, namely, when the liquid fraction in the PCM volume corresponded to 10% (1050 s), 50% (2800 s), 80% (4120 s), and 100% (6330 s), where the five

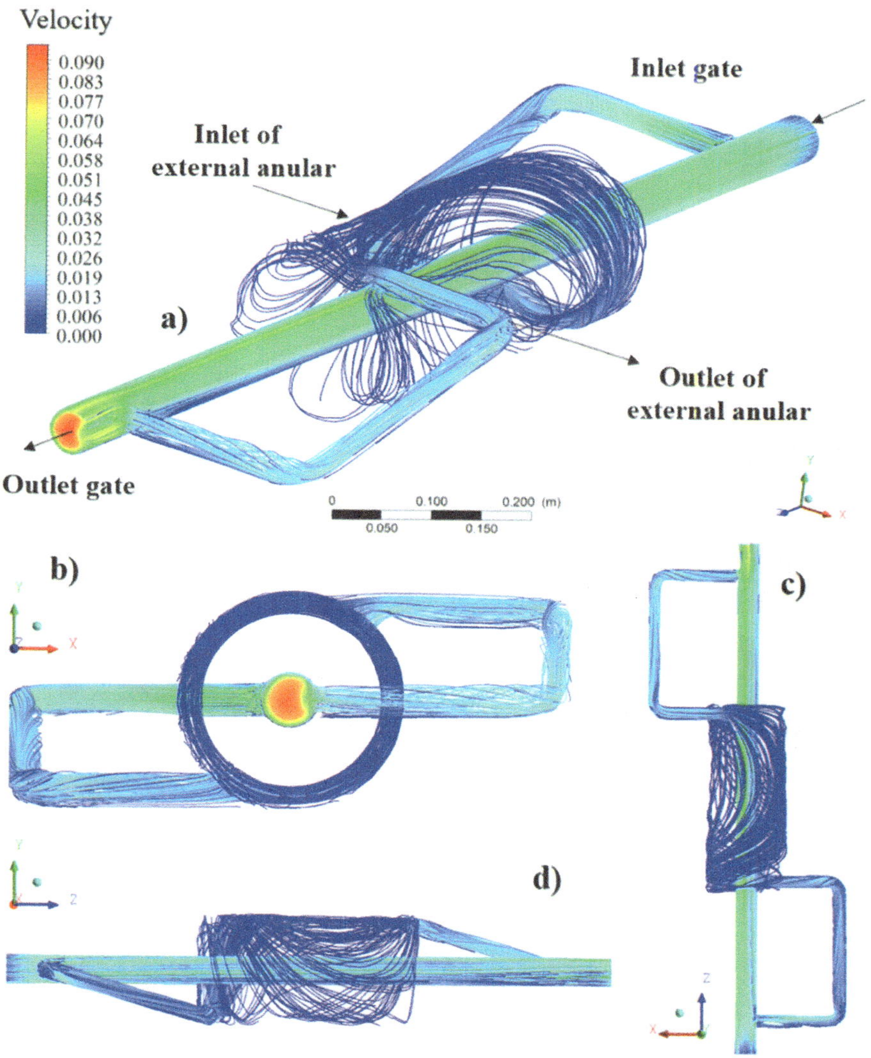

Fig. 15 Streamlines with velocity levels in **a** isometric, **b** front, **c** top, and **d** lateral views

section planes already specified were analyzed. In addition, the isometric view of the entire physical domain is presented.

First, it was possible to observe the expected physical coherence in the heat exchanger: the HTF entered at 90 °C at the external and internal entrances of the TTHX and left the heat exchanger with reduced temperatures as it released thermal energy to the PCM, which gradually increased its temperature until it was completely melted. Along the length of the TTHX, the HTF reduced its temperature, losing

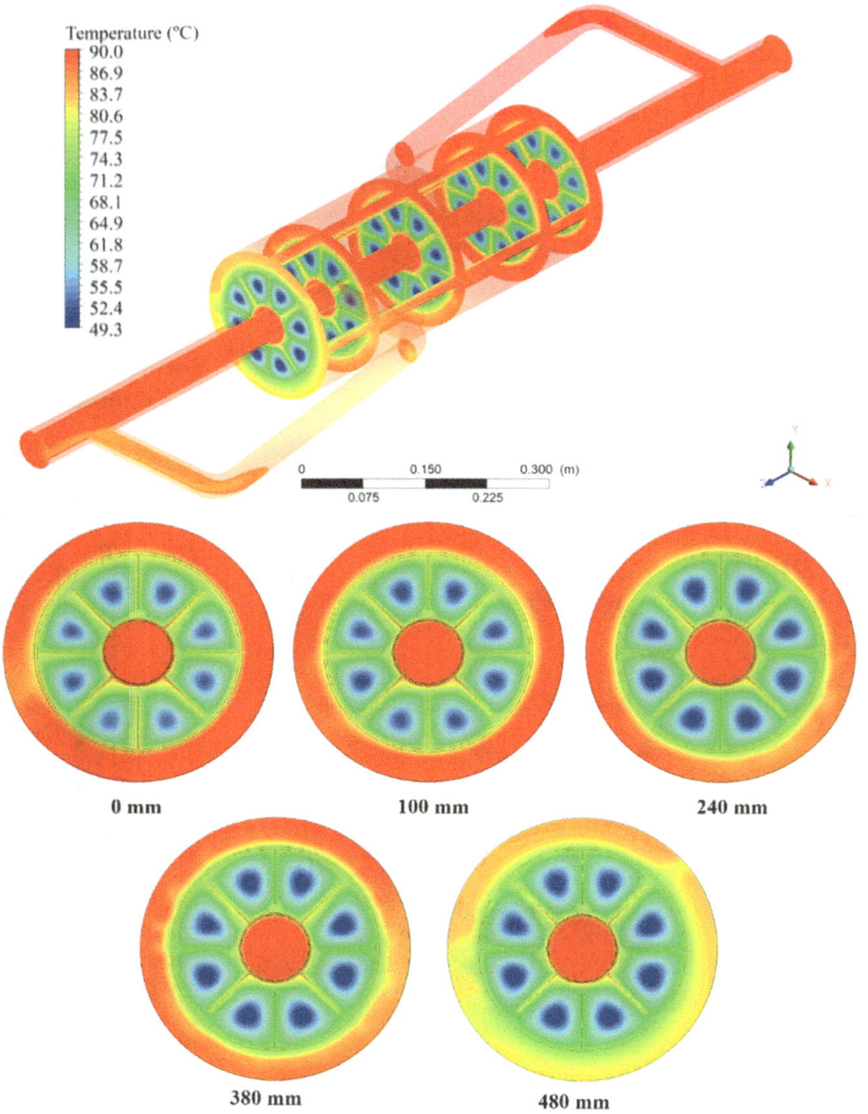

Fig. 16 Temperature distribution in the TTHX domain for 10% melted PCM volume (1050 s)

energy to the inlet sections and heating the sections closest to the outlet of the heat exchanger more slowly, which were the last to melt.

In Fig. 16, it was observed that the HTF, which entered the exchanger at 90 °C, left the external region of the domain with temperatures around 80 °C. In the internal region of the HTF, the temperatures in the regions very close to the internal surface of

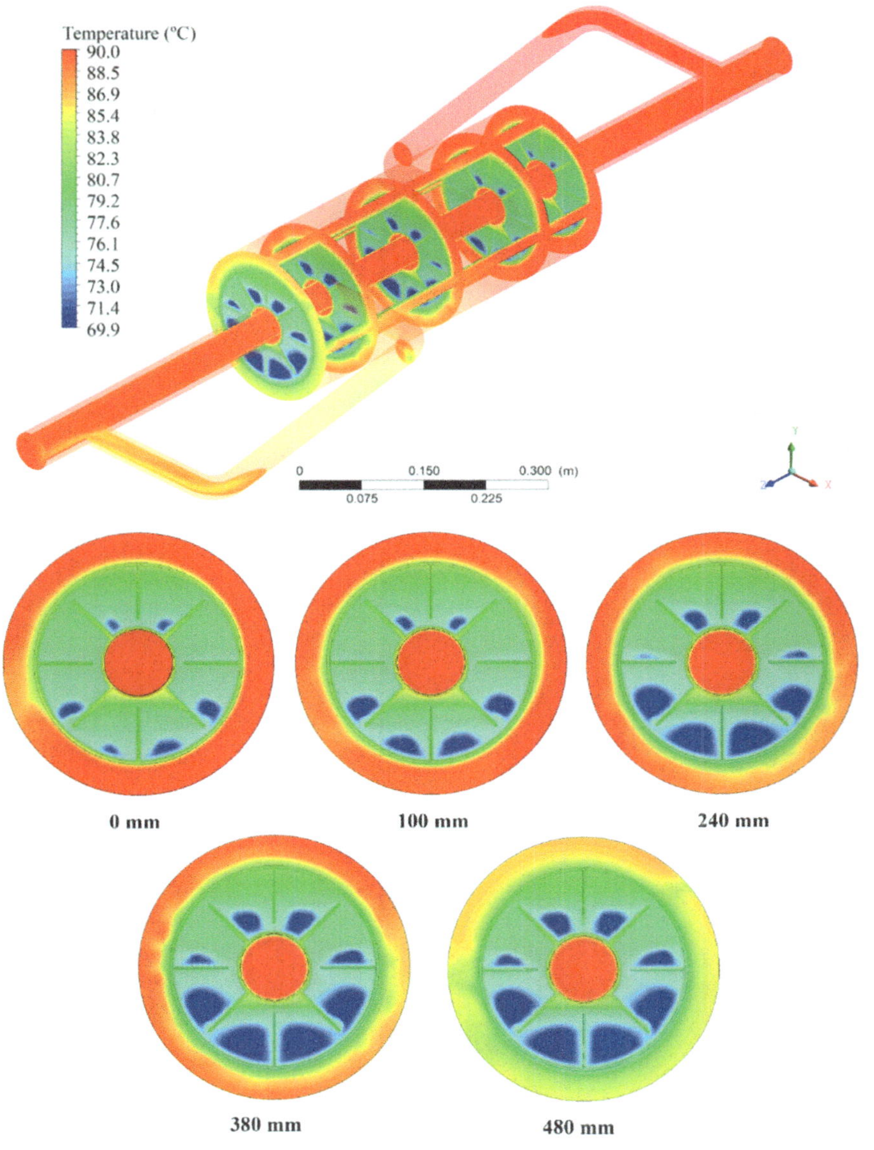

Fig. 17 Temperature distribution in the TTHX domain for 50% melted PCM volume (2800 s)

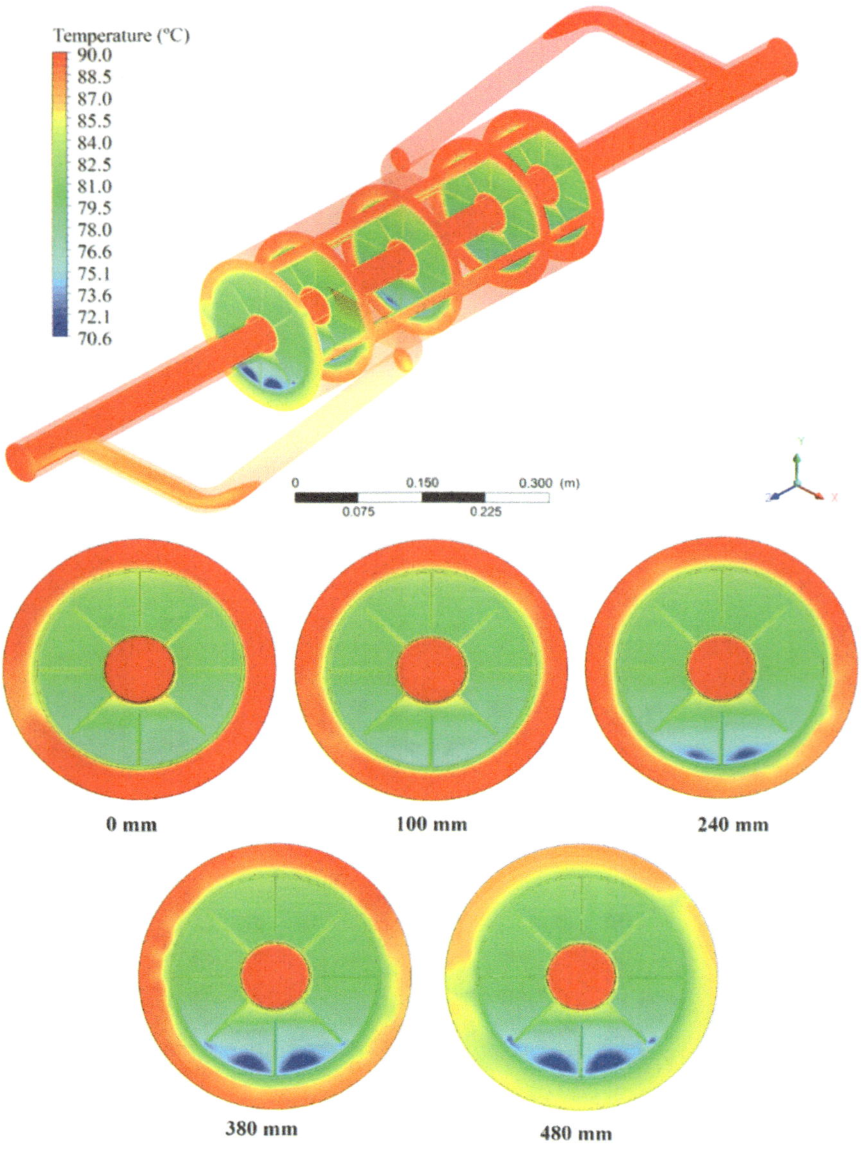

Fig. 18 Temperature distribution in the TTHX domain for 80% melted PCM volume (4210 s)

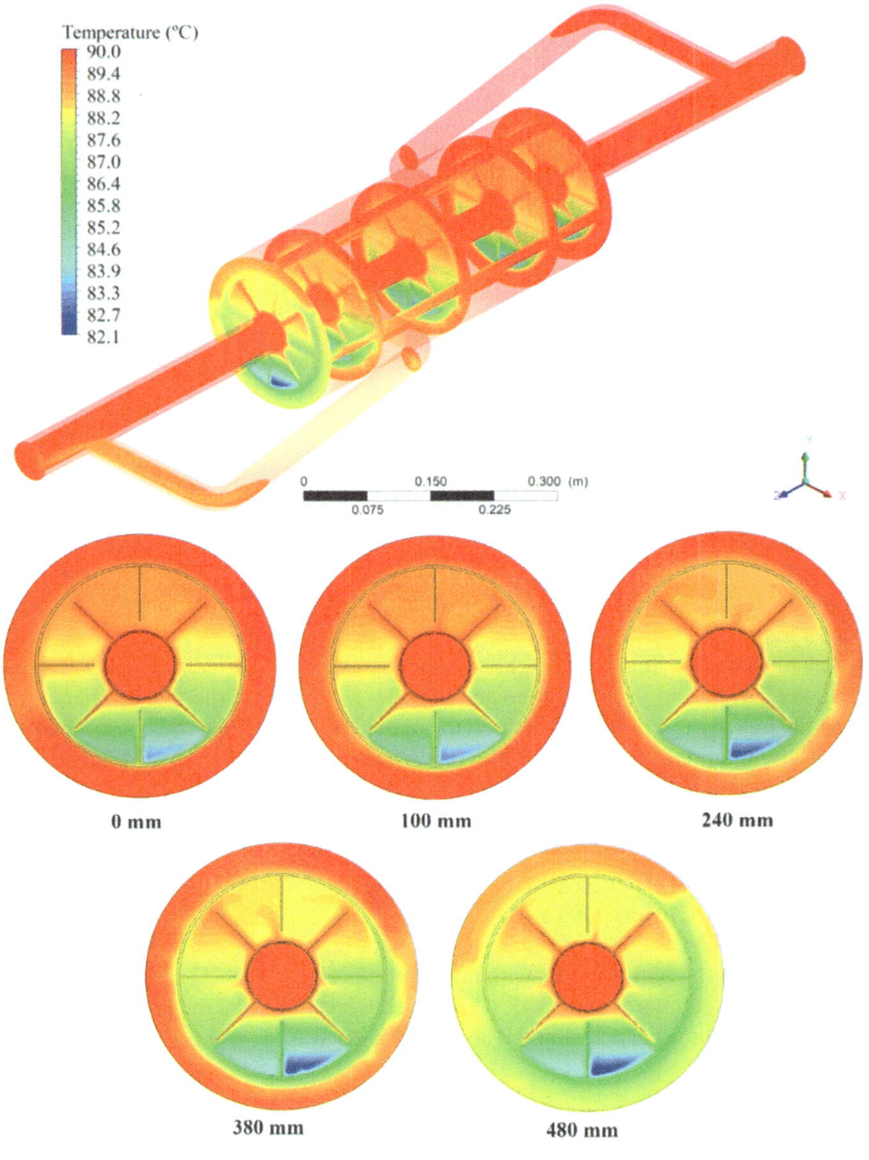

Fig. 19 Temperature distribution in the TTHX domain for 100% melted PCM volume (6330 s)

the inner tube were reduced from 90 °C to around 85 °C, indicating that the thermal boundary layer in this region was very thin. Furthermore, it can be seen that there were two important mechanisms of heat transfer acting in the TTHX: in the external region, there was a larger thermal exchange area and a longer contact time between the HTF and the pipe wall, which caused the heat exchange fluid to come out colder from the TTHX external annular; in the internal part, there was a smaller contact area but higher fluid velocity levels, which made the heat transfer via convection from the internal surface of the PCM annular more intense.

Even in Fig. 16 of the PCM domain, the regions close to the fins heated up more quickly due to the high copper thermal conductivity. It was observed that in the sections located at $z = 0$ and 480 mm, the temperatures in the regions close to the fins were close to 80 °C. The PCM domain had a temperature range from 80 to 49.3 °C in the sections at $z = 240$, 380, and 480 mm, and from 80 to 55 °C in the sections at $z = 0$ and 10 mm.

In Fig. 17, it is possible to observe the HTF left more heated, at around 85 °C, from the TTHX external annular. Since the PCM was heated, the temperature gradient between it and the HTF was reduced, and thus, the driving force for the heat transfer was also reduced. Regarding the PCM, it appears that as it moved through $z = 100$, 240, 380, and 480 mm, fractions with lower temperatures were accumulated in the annular lower region. This was due to the PCM density reductions, which were related to the increase in paraffin temperature.

The most heated fluids were lighter and tended to be deposited in the regions near the lower surfaces of the inner tube and the upper surfaces of the intermediate tube, while the heavier fractions collected on the upper face of the inner tube and the lower face of the intermediate tube; the fins' geometry restricted the upward movements of the lighter PCM fractions within the annular.

For an 80% melted PCM volume (Fig. 18), HTF was removed from the external annular and the internal tube with temperatures close to 87 °C. For the PCM, it was possible to observe regions close to the fins with temperature levels above the material melting temperature (82.17 °C).

When the PCM had 100% of its volume melted (Fig. 19), the HTF left the external and internal region of the TTHX with temperatures close to 88 °C and 90 °C, respectively, with all sections above 82.17 °C. It was observed that the total processing time to reach 100% liquid material in the heat exchanger was around 6330 s (105.5 min).

Liquid Fraction Fields

Figures 20, 21, 22, 23 and 24 show the distribution of the liquid fractions throughout the melting process with 10% (1050 s), 30% (1990s), 50% (2800 s), 70% (3650 s) and 90% (4690 s) of the volume of the PCM being melted, respectively.

In the step referring to Fig. 20, with a range of a 0 to 100% liquid fraction, the following characteristics were observed: In the $z = 0$ and 100 mm sections, the PCM fractions near the fins were found with values around 50%; in the regions referring to the lower half of the annular and close to the external surface of the inner tube,

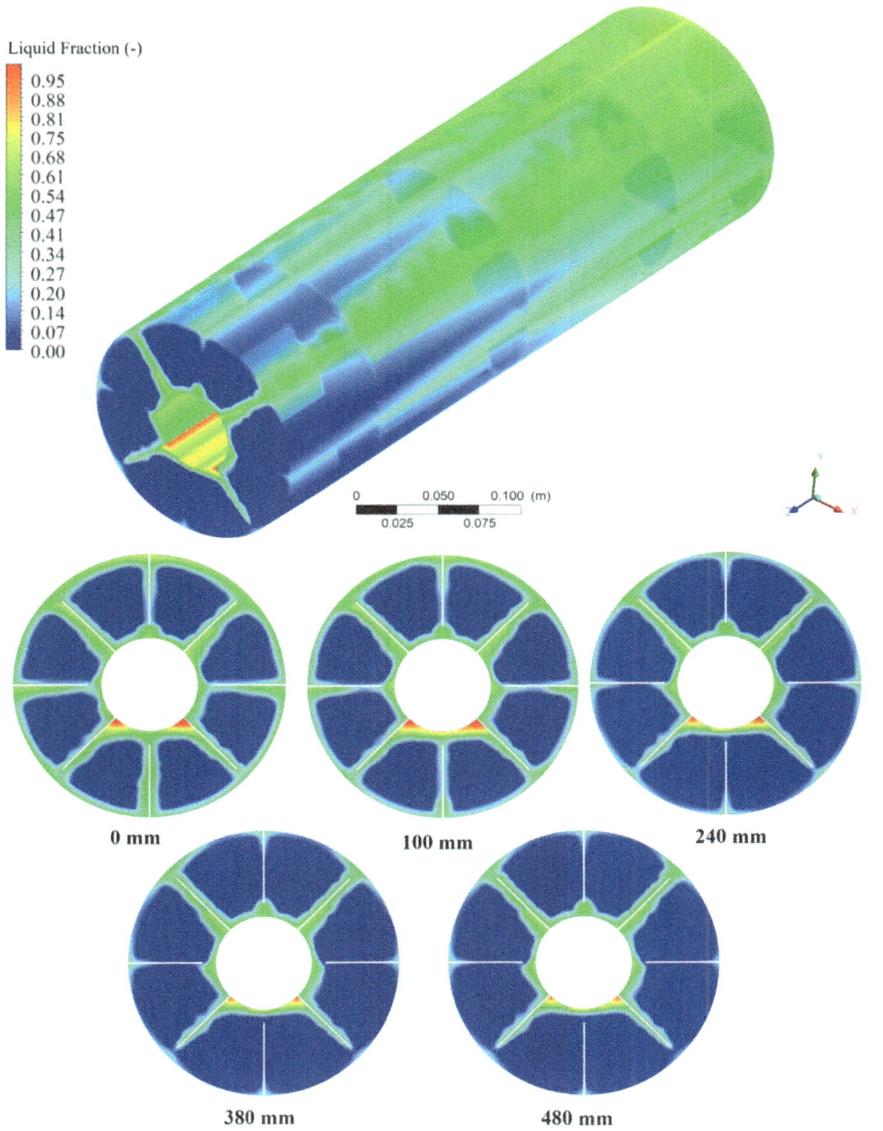

Fig. 20 Distribution of the liquid fraction of PCM at t = 1050 s (10% of its volume melted)

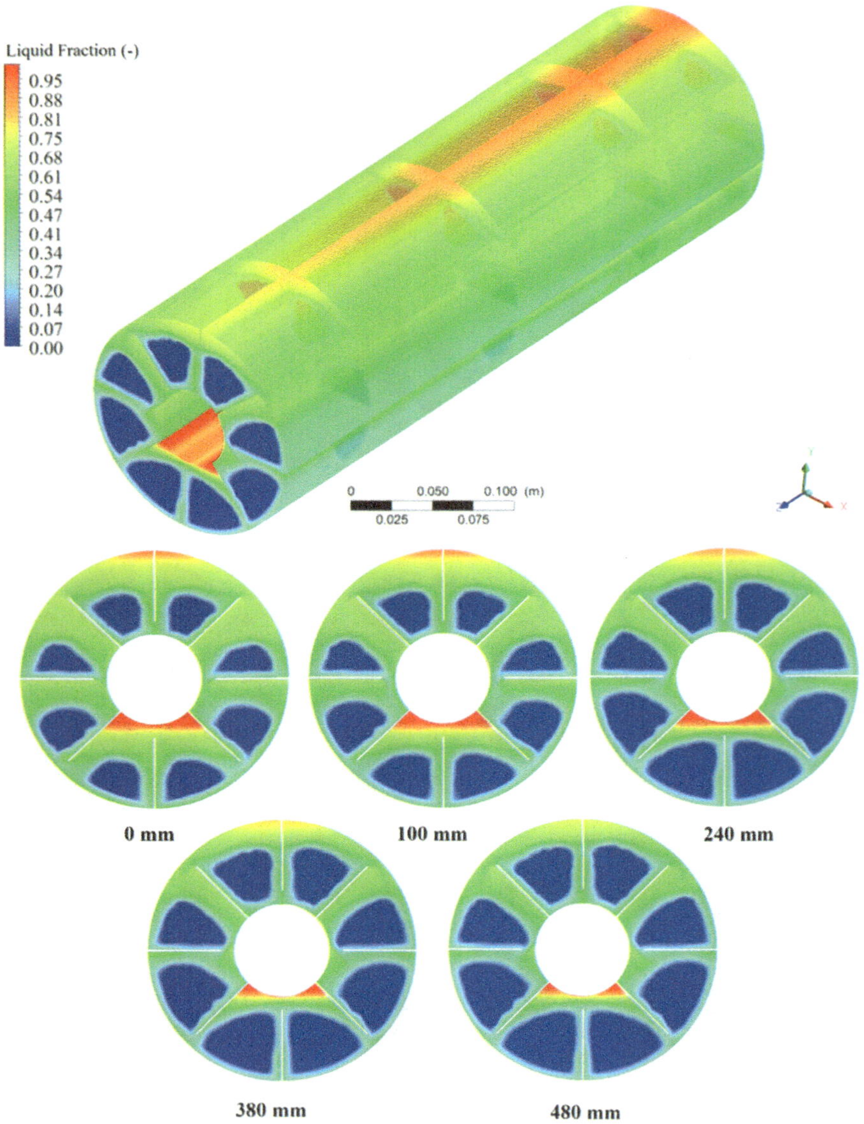

Fig. 21 Distribution of the liquid fraction of PCM at t = 1990s (30% of its volume melted)

the local liquid fraction values close to 100% were obtained. This was due to the high heat flux from the TTHX internal tube. At the $z = 240$ mm section, there was a reduction in the material's melting intensity, with reduced portions of the material undergoing a phase change in these regions close to the annular's external surface. This occurred in connection with the cooling of the HTF along the TTHX's length. At

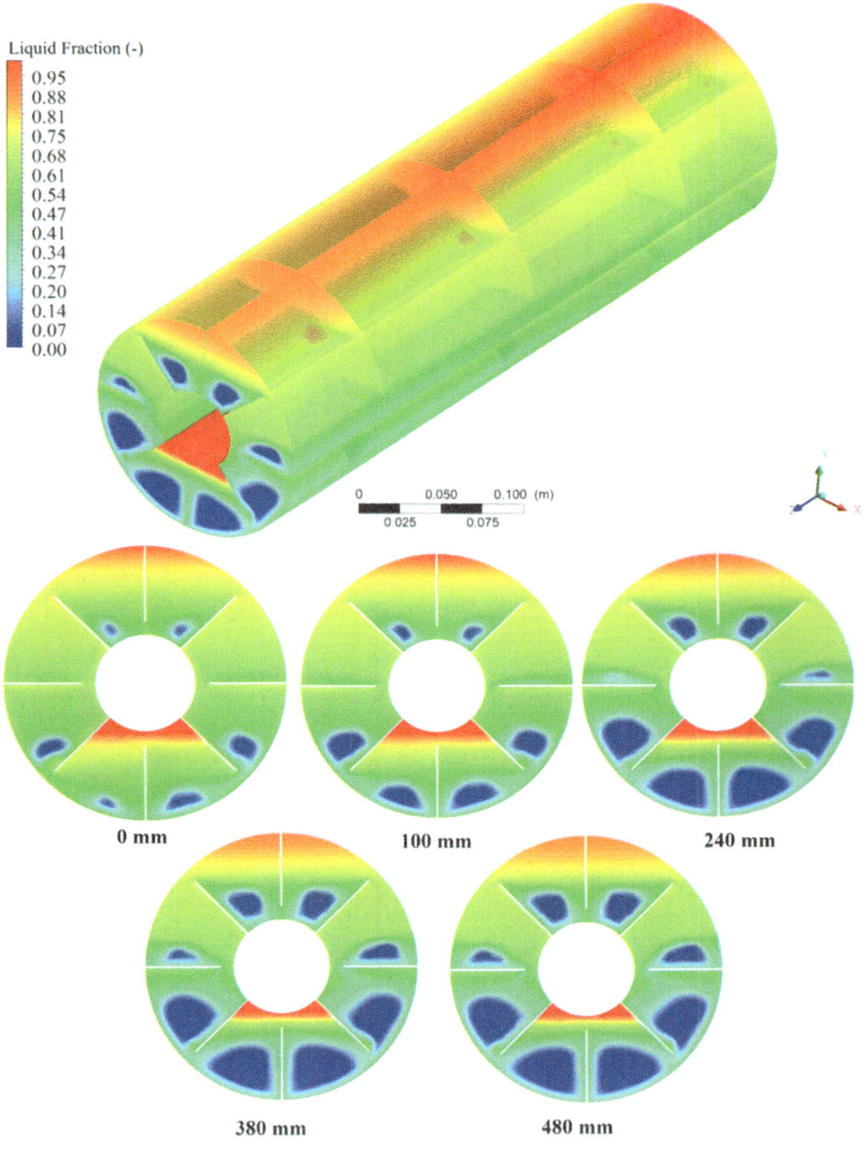

Fig. 22 Distribution of the liquid fraction of PCM at t = 2800 s (50% of its volume melted)

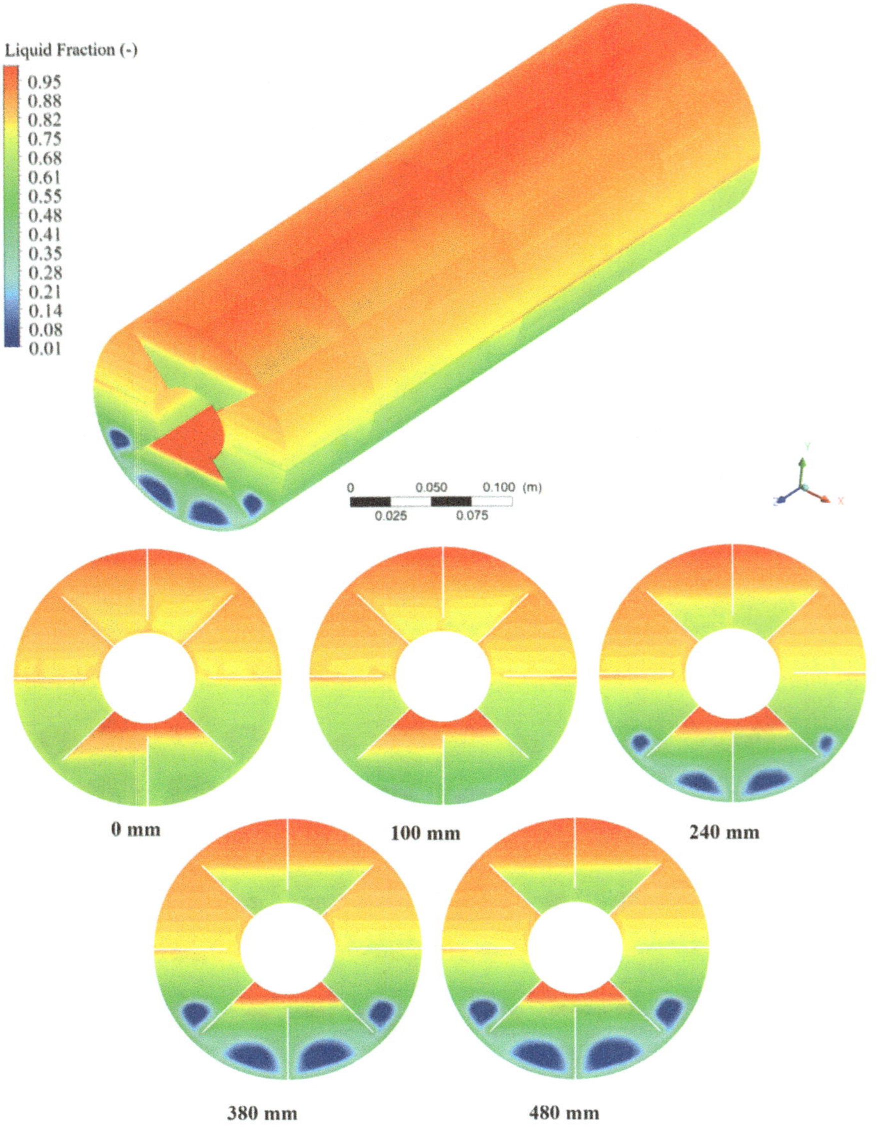

Fig. 23 Distribution of the liquid fraction of PCM at t = 3650 s (70% of its volume melted)

the sections located at $z = 380$ and 480 mm, it was observed that the liquid fractions with values close to 50% were restricted to the regions close to TTHX internal tube.

In Fig. 21, the following characteristics can be verified, with a range of 0 to 100% liquid fraction: The TTHX sections measured at $z = 0$ mm, 100 mm, and 240 mm presented very similar liquid fractions distributions; however, reductions in

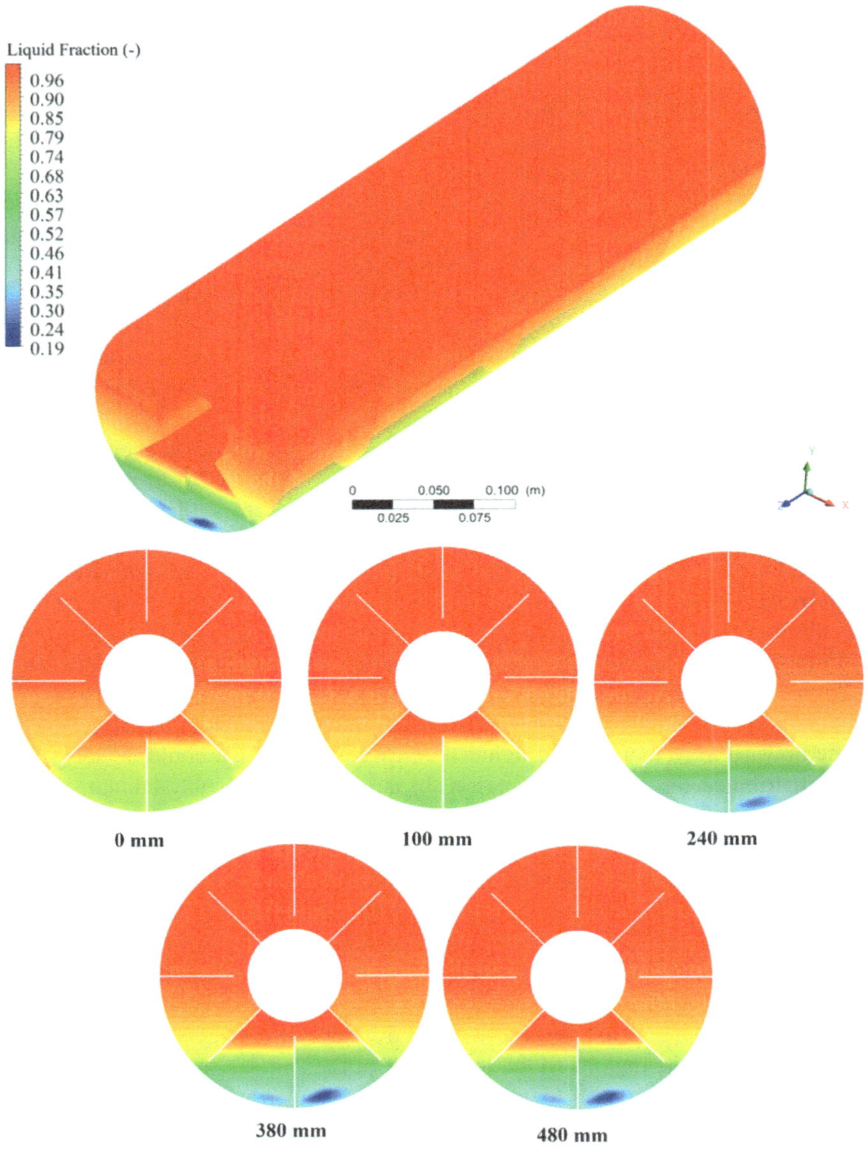

Fig. 24 Distribution of the liquid fraction of PCM at t = 4690 s (90% of its volume melted)

the melting intensity from the section $z = 0$ mm and a more heated, and therefore, lighter material accumulation was observed in the upper regions of the lower and upper half of the annular where the PCM was located. The reduction in PCM density caused the accumulation of heated fractions in the upper regions and retention in the lower half due to the fin's geometry. In the sections at $z = 380$ and $z = 480$ mm, most of the PCM was still in a solid state.

From the analysis of Fig. 22, the following characteristics can be observed, with a range of 0 to 100% liquid fraction: At the entrance of the TTHX ($z = 0$ mm), almost all the material was already in the transition region, with values between 60 and 100% in a liquid state, except for two small localized regions that had liquid fractions that were still close to 10%. In all sections, the fractions were completely melted at the bottom of the annular inner surface, the top of the annular external surface had values close to 75% at $z = 380$ and 480 mm, and the fraction was close to 100% in the sections $z = 0$, 100, and 240 mm. At the $z = 100$ mm section, heavier fractions were found completely in a solid state, which had accumulated at the annular lower part due to variations in densities of the PCM. In the section at $z = 240$ mm, the accumulation of PCM fractions that were still in the solid state occurred in the regions close to the annular inner surface's upper part and had values below 20% in the regions close to the annular external surface's lower part. At the sections located in $z = 380$ and 480 mm, the same accumulation of solid material was observed to occur more intensely, even in the annular upper half.

In Fig. 23, the following characteristics can be seen within a range of 1 to 100% of liquid fraction: At the entrance of the TTHX ($z = 0$ mm), almost the entire upper half of the annular was found with liquid fractions above 88%. This was due to the upward movement of the lighter PCM fractions, which could be seen to be more distributed throughout this region. In the lower half of the annular, a more homogeneous distribution of liquid fractions was observed, with values close to 70% liquid composition, except for the regions close to the annular internal surface's lower part, which retained material fractions that had already completely melted due to the fin's geometry. The differences between the liquid fraction distributions in the PCM in the x–y planes at $z = 0$ and 100 mm were very small, indicating that the process tended to occur uniformly in the sections along the length as the melting process approached the end.

In the sections at $z = 240$, 380 and 480 mm from the entrance, 10% liquid fractions were observed in the regions of the lower half of the annular, and similarly for the three sections, in the central region of the annular upper half, the liquid fractions were around 80%.

In Fig. 24, the distributions of liquid fractions ranged from 19 to 100%. The following characteristics can be seen: In all the analyzed sections, the entire upper half of the annular was found with liquid fractions around 100%. In the sections at $z = 240$, 380, and 480 mm, there was a tendency to heat the material contained in the annular left part slightly more intensively than in the right part. This phenomenon was more easily observed at this process stage and was due to the swirling movement of the HTF that developed in the heat exchanger along the external annular of the HTF, which for regions close to the section at $z = 480$ mm, presented greater velocities in the lower left part of the TTHX.

Velocity and Pressure Distribution in the PCM

In the distributions illustrated in Fig. 25, it can be seen that velocity fields developed in the PCM due to the buoyant forces in the material structure, which was a consequence of the density variations that the PCM heating provides. The velocity levels and trajectory of the fluid throughout the process are indicated by vectors.

The PCM started the melting process in the fully solidified state at zero velocity. As the material reached a temperature level higher than the melting temperature, it reduced its density, which under the force of gravity, began moving according to the degree of the material's melting stage. These movements allowed for the heat transfer process, which also occurred via natural convection instead of being purely diffusive.

When 10% of the PCM volume was in the liquid state, non-zero velocities were located close to the inner surface of the PCM annular and decreased from the section at $z = 100$ mm to the section at $z = 380$ mm, where there was no material in the solid state.

When the PCM had 50% of its volume in the liquid state, it was verified that the velocities rose to a maximum value of 4.85×10^{-4} m/s near the fins. It was observed that the movement of the PCM fractions initially occurred randomly. At $z = 240$ mm, there was a greater distribution of non-zero velocities due to the phase transition that caused the movement of the PCM fractions to be more intense in this section.

When 80% of the material was in the liquid state, the velocities were higher at around 1×10^{-3} m/s, with these having developed in the annular upper half. On this occasion, the PCM fraction velocities were more distributed at $z = 100$ and 240 mm, while at $z = 380$ mm, it was possible to verify intense movement being developed near the fins.

When the PCM had 100% of its volume in the liquid state, the velocities, although more distributed, tended to be reduced given that the driving force of the movements was the density variations, which had ceased to occur. It was observed that the movement of the PCM fractions presented a more ordered behavior due to the material being in a completely liquid state.

Figure 26 illustrates the pressure distributions at $z = 240$ mm for 10%, 50%, 80%, and 100% of the material in the liquid state. There were no significant differences in the pressure fields for the other sections along the annular's length. Due to the reduction in the density of the material, throughout the melting process, increases in pressure on the PCM volume were observed. Initially, with the material fully solidified, pressure variations along the volume were non-existent; however, as the material underwent the melting process, it tended to expand the volume, which was restricted by the walls of the TTHX.

Due to the accumulation of liquid material in the upper part of the annular, the maximum pressure values were observed in this region. Meanwhile, in the lower part,

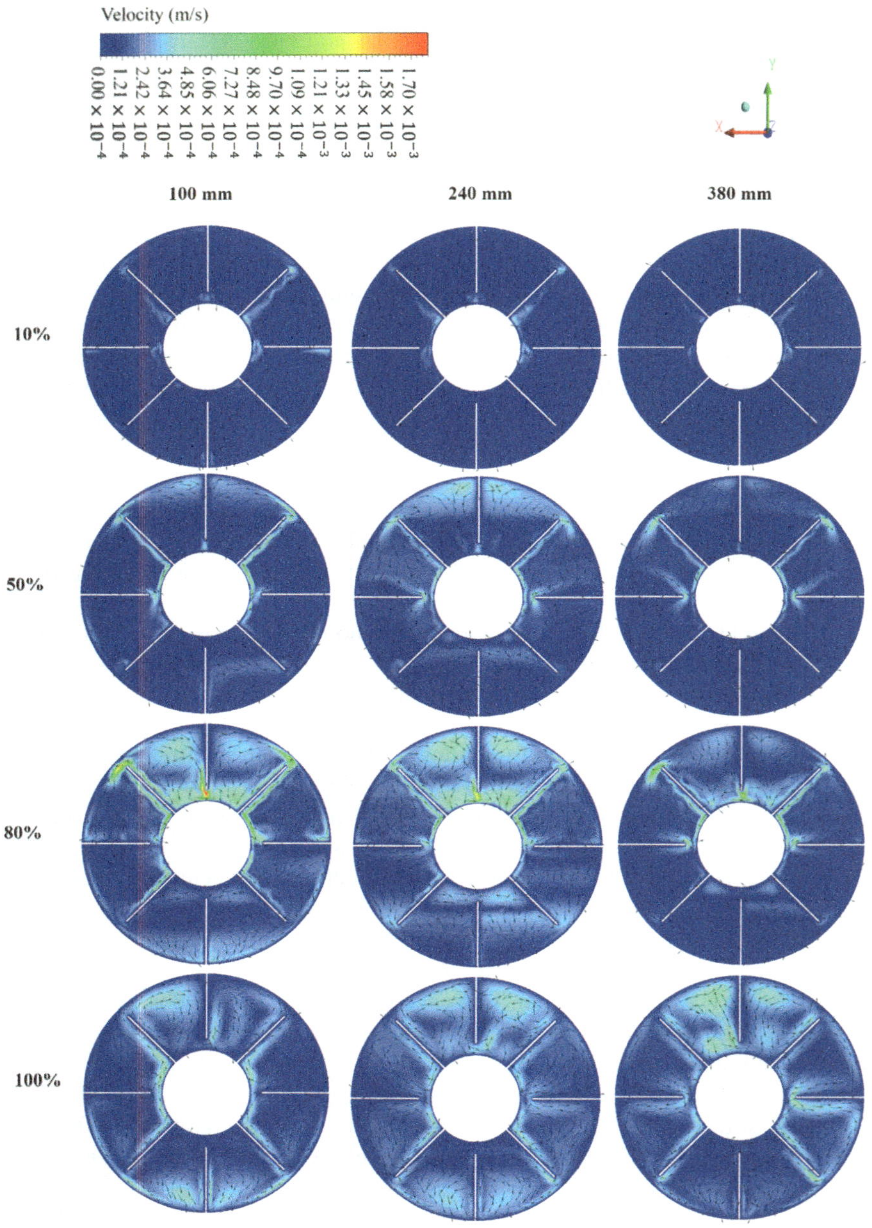

Fig. 25 Velocity distribution in the PCM domain at different fractions of its volume being melted

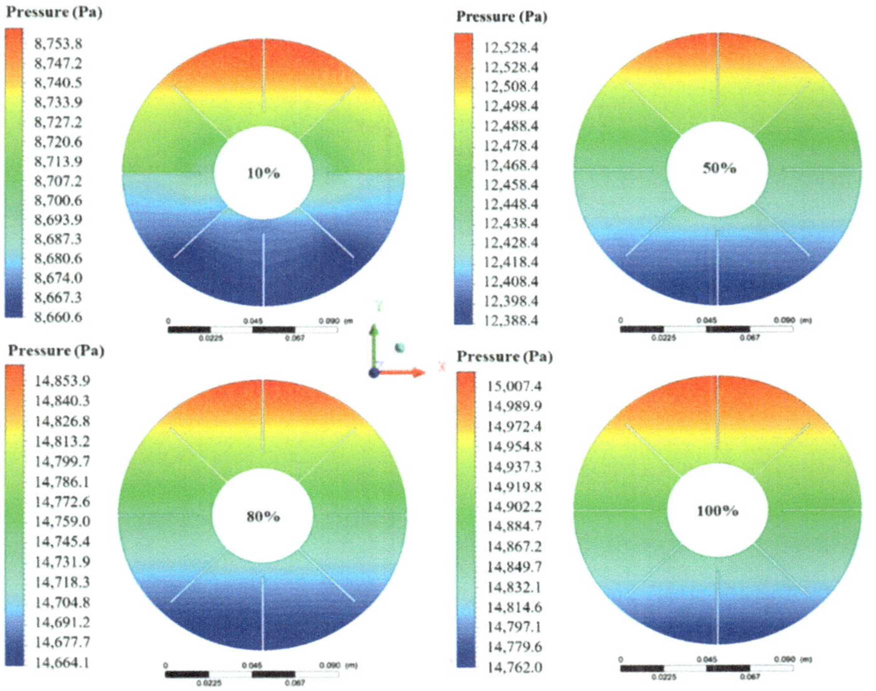

Fig. 26 Pressure distributions in the PCM domain with different fractions of its volume being melted at z = 240 mm

which was the last region to completely melt, had the lowest pressure levels. With 10% of the PCM in the liquid state, the manometric pressures were verified to reach maximum values around 8760 Pa and had a minimum of 8660 Pa, while the center line of the annular had intermediate values. When the PCM had 50% of its volume in the liquid state, the maximum pressure levels rose to 12,538.4 Pa and had a minimum of 12,338.4 Pa, and due to the increase in the liquid fraction, there was a greater pressure distribution in the central regions of the annular. The pressure fields corresponding to the liquid fractions of 80 and 100% were qualitatively and quantitatively very similar, with maximum levels at around 15,000 Pa and had minimum levels of 14,760 Pa. This occurred due to the progress of the melting process in the entire volume of the PCM.

Furthermore, it is important to state that the local parameters' behavior was affected by the volumetric thermal and fluid-dynamic parameters' behavior.

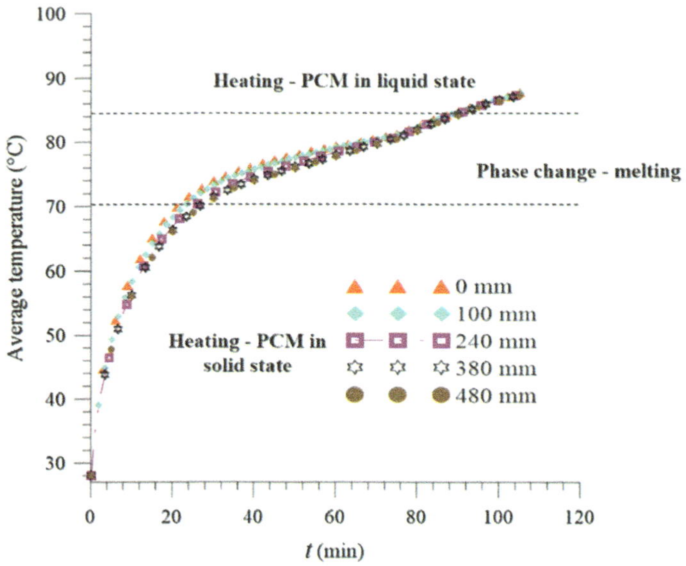

Fig. 27 Transient behavior of the PCM's average temperature

Analysis of the Averages Heat and Mass Transport Variables

Figure 27 illustrates the average temperatures measured in the PCM (average at the surface) in different sections ($z = 0$, 100, 240, 380, and 480 mm) during the melting process, with the HTF at 90 °C and the PCM initially at 27 °C. The graph highlights three distinct stages of the melting process: (1) heating the solid PCM to the material melting temperature of 70.12 °C, (2) the phase transition that occurred between 70.12 and 82.17 °C, and (3) the heating of the liquid PCM to average temperature values above 82.17 °C. This last step represented the time necessary to melt the PCM contained in the volume.

The first stage occurred within approximately 21.5 min for the entry section $z =$ 0 mm and around 27.16 min for section $z = 480$ mm. The second stage occurred between 20 and 81 min, with maximum variations between the $z = 0$ and 480 mm sections of around 3 °C, which occurred in approximately $t = 40$ min. The third stage took between 80 and 105.5 min, with very close values for all sections. As previously described, these differences in the average temperatures in the specified plans occurred due to the heating of the PCM along its length simultaneously to the cooling of the HTF, which caused the sections distant from $z = 0$ mm to heat up more slowly.

In Fig. 28, the transient measurements of the average temperature of the HTF throughout the melting process are presented, taken at section $z = 880$ mm. Initially, the water that entered the pipes with a temperature of 90 °C was abruptly cooled to values close to 87 °C. Then, it started to increase its temperature up to 88 °C within

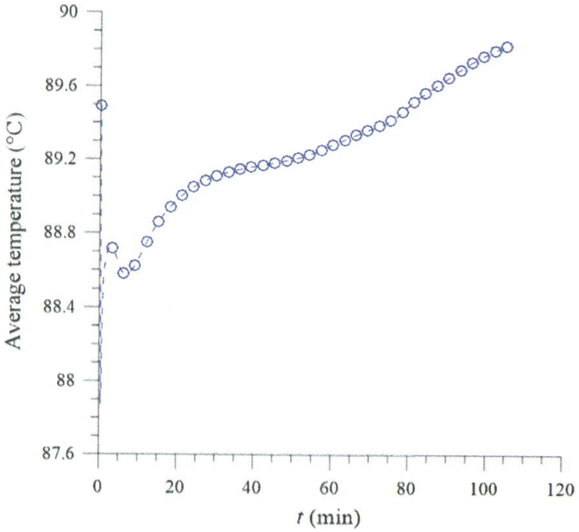

Fig. 28 Transient behavior of the average fluid temperature at the heat exchanger outlet

the first 10 min of heating, which corresponded to the time taken for the average temperatures of the PCM sections to reach its liquid temperature. Between 10 and 20 min, the HTF was heated quickly to 88.9 °C, and subsequently, continued to increase its temperature with an almost constant rate until the end of the process, where it reached values close to 89.75 °C. The heat flux was reduced over time because the temperature differences between the PCM and HTF were reduced during the melting process.

Figure 29 illustrates the results obtained for the average heat fluxes, measured at the inner surface of the internal tube and the external surface of the intermediate tube during the melting process. These results indicated the amounts of energy associated with the HTF flow transferring heat to the PCM. From the analysis of this figure, it was observed that a peak of heat flux occurred at the beginning of the PCM loading process for the two referred surfaces due to the thermal shock between the HTF and PCM temperatures. For the inner surface of the internal tube, the peak flux was approximately 12.5 kW/m^2, while the peak flux was 16 kW/m^2 for the outer surface of the external tube. This difference was due to the variation in velocity between the internal and external annular tubes. As the PCM was heated, the intensity of the heat fluxes on both surfaces was reduced. It was observed that, within approximately 15 min, the average heat flux on the external surface was close to 1 kW/m^2, while on the internal surface, it was about 3 times this value, which was close to 3 kW/m^2. Throughout the melting process, these heat fluxes were reduced at an approximately constant rate until the end of the melting period, where they reached values close to zero.

Figure 30 presents the average liquid fractions measured at the sections $z = 0$, 100, 240, 380, and 480 mm from the TTHX inlet during the PCM loading process. It was observed that it took about 8 min for the most distant section, $z = 480$ mm, to

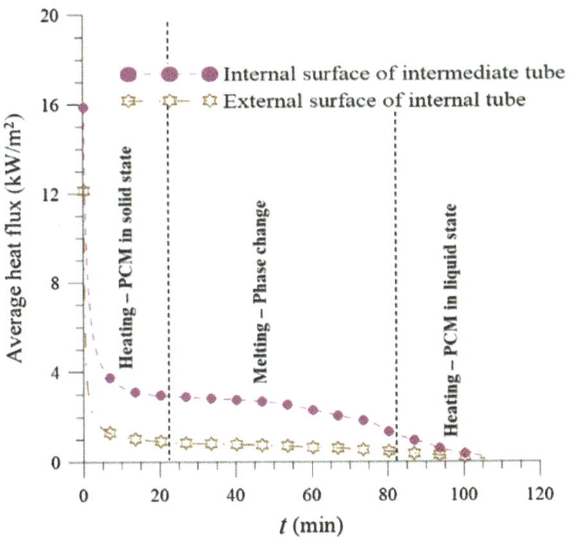

Fig. 29 Transient behavior of the heat flux to the outer surfaces of the intermediate tube and the inner surface of the inner tube

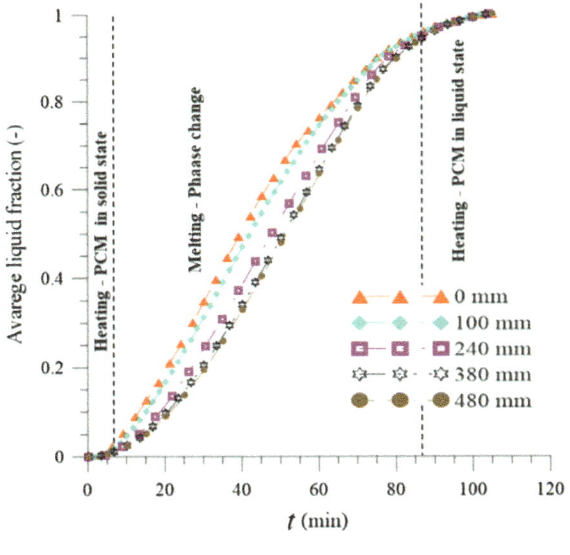

Fig. 30 Transient behavior of the average liquid fraction for different sections of the PCM domain

reach non-zero liquid fraction values after the material entered the transition region. During the loading, there were few differences between the sections at $z = 380$ and 480 mm from the entrance, which were brought to the state of an 80% liquid fraction in about 86 min. Subsequently, the increase in liquid fraction occurred more slowly, where the material reached a 100% liquid fraction at 105.5 min.

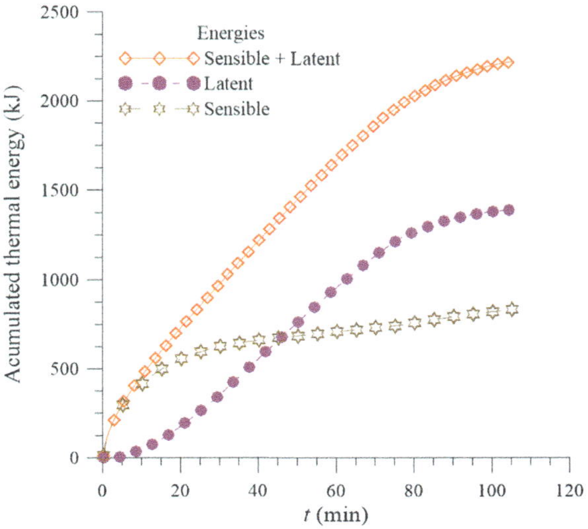

Fig. 31 Transient behavior of the thermal energy absorbed (volumetric) by the PCM during the melting process

Figure 31 presents the energy absorption levels associated with the 6.89 kg of PCM subjected to the melting process. By analyzing this figure, it can be seen that there was an absorption of sensible energy, which was predominant for the first approximately 10 min of TTHX operation; the material took around 5 min to start the phase transformations and start absorbing latent thermal energy. Before that, the only form of energy absorption was via sensible heat, which reached values close to 500 kJ within this timeframe. Therefore, the curves referring to the sum of the absorbed thermal energy via the latent and sensible heat and sensible heat alone were almost identical up to this point. Subsequently, the latent energy absorption increased until it exceeded the sensible heat curve at 46 min (677 kJ) of melting process; then, the latent heat curve increased until 1330 kJ due to the phase change and the sensible heat curve reaches 835 kJ due to the increase in the temperature to values above the melting point (82.17 °C) by the material in the liquid state. At the end of 105 min, the complete melting of the material was achieved and a total of 2220 kJ of thermal energy had accumulated.

4 Concluding Remarks

In this chapter, the physical problem of the melting process of a phase change material (PCM) in triplex tube heat exchanger (TTHX) with finned copper tubes has been studied. Because the great importance, emphasis is given to the heat transfer between a heating fluid (HTF) and phase change material to building application (liquid-desiccant air conditioning system). All the research was developed using computational fluid dynamic as analysis tool.

From the obtained results, the following conclusions can be drawn: (a) The proposed mathematical model was able to predict the transient behavior of the PCM during the melting process; (b) As the phase change material heated up along the length of the TTHX and due to the corresponding water (HTF) cooling that occurred, the PCM heating process was higher in the sections next to $z = 0$ mm and lost power over the TTHX's length; (c) the increase in the temperatures of the PCM and the resulting reductions in density moved of the lighter fractions to the upper parts of the PCM annular and increase in the pressure inside the PCM volume; (d) The average heat fluxes through the internal surface of the inner tube were about 3 times the heat fluxes through the outer surface of the intermediate tube, which occurred because the velocities in the inner tube were about 10 times higher than those of the TTHX's external annular, and (e) The latent energy accumulated during the melting process was 1330 kJ, while the accumulated sensitive energy was 835 kJ.

Acknowledgements The authors thank to CNPq, CAPES and FINEP (Brazilian Research Agencies) for the financial support, and the scientific support from the researchers mentioned along this chapter.

References

Abhat A (1983) Low temperature latent heat thermal energy storage: heat storage materials. Sol Energy 30(4):313–332

Abdulateef AM, Mat S, Abdulateef J, Sopian K, Al-Abidi AA (2018) Geometric and design parameters of fins employed for enhancing thermal energy storage systems: a review. Renew Sustain Energy Rev 82:1620–1635

Al-Abidi AA, Mat S, Sopian K, Sulaiman MY, Mohammad AT (2013) Numerical study of PCM solidification in a triplex tube heat exchanger with internal and external fins. Int J Heat Mass Transf 61:684–695

Al-Abidi AA, Mat S, Sopian K, Sulaiman MY, Mohammad AT (2014) Experimental study of melting and solidification of PCM in a triplex tube heat exchanger with fins. Energy Build 68:33–41

Antunes PF (2016) Numerical study of thermal energy storage units with phase change materials for the thermoregulation of photovoltaic panels. Master´s dissertation, Mechanical Engineering, Coimbra University, Coimbra, Portugal (in Portuguese)

Ansys Inc. (2015) ANSYS FLUENT Theory Guide; Release 15.0; ANSYS Inc., Canonsburg, Pennsylvania, USA, 814 p

Anderson WK, Bonhaus DL (1994) An implicit upwind algorithm for computing turbulent flows on unstructured grids. Comput Fluids 23:1–21

Bahrani SA, Royon L, Abou B, Osipian R, Azzouz K, Bontemps A (2017) A phenomenological approach of solidification of polymeric phase change materials. J Appl Phys 2017:121

Barth T, Jespersen D (1989) The design and application of upwind schemes on unstructured meshes. In: Proceedings of the 27th aerospace sciences meeting, Reno, Nevada, 9–12 January 366

Costa S (2014) Thermal energy storage through phase change materials, Master's Dissertation, Mechanical Engineering, University of Porto, Porto, Portugal (in portuguese)

Cabeza L, Mehling H (2008) Heat and cold storage with PCM. An up to date introduction into basics and applications. Springer, Berlin

Esapour M, Hosseini SMJ, Ranjbar AA, Pahamli Y, Bahrampoury R (2016) Phase change in multitube heat exchangers. Renew Energy 85:1017–1025

García-Valladares O (2004) Numerical simulation of triple concentric-tube heat exchangers. Int J Thermal Sci 43(10):979–991
Himran S, Suwono A, Mansoori GA (1994) Characterization of alkanes and paraffin waxes for application as phase change energy storage medium. Energy Sources 16(1):117–128
Hasnain SM (1998) Review on sustainable thermal energy storage technologies. Part I: Heat storage materials and techniques. Energy Conversion Manag 39(11):1127–1138
Ibrahim NI, Al-sulaiman FA, Rahman S, Yilbas BS, Sahin AZ (2017) Heat transfer enhancement of phase change materials for thermal energy storage applications: a critical review. Renew Sustain Energy Rev 74:26–50
Jmal I, Baccar M (2015) Numerical study of PCM solidification in a finned tube thermal storage including natural convection. Appl Therm Eng 84:320–330
Karthikeyan S, Ravikumar Solomon G, Kumaresan V, Velraj R (2014) Parametric studies on packed bed storage unit filled with PCM encapsulated spherical containers for low temperature solar air heating applications. Energy Conversion Manag 78:74–80
Kenisarin M, Mahkamov K (2016) Passive thermal control in residential buildings using phase change materials. Renew Sustain Energy Rev 55:371–398
Kośny J (2015) Short history of PCM applications in building envelopes. In: Kośny J (ed) PCM-Enhanced Building Components: An Application of Phase Change Materials in Building Envelopes and Internal Structures. Springer, Cham, Switzerland, pp 21–59
Kuznik F, David D, Johannes K, Roux JJ (2011) A review on phase change materials integrated in building walls. Renew Sustain Energy Rev 15:379–391
Lane GA (1983) Solar heat storage: latent heat materials. Background and scientific principles. 1, CRC Press, Boca Raton
Launder BE, Spalding DB (1972) Mathematical models of turbulence. Academic Press, New York
Leonard BP, Mokhtari S (1990) ULTRA-SHARP Nonoscillatory Convection Schemes for High-Speed Steady Multidimensional Flow. Technical Report Number NASA TM-102568, ICOMP-90–12; NASA: Cleveland, Ohio, USA (1990) 54 p
Li WQ, Qu ZG, He YL, Tao WQ (2012) Experimental and numerical studies on melting phase change heat transfer in open-cell metallic foams filled with paraffin. Appl Thermal Eng 37:1–9
Li Z, Wu ZG (2015) Analysis of HTFs, PCMs and fins effects on the thermal performance of shell-tube thermal energy storage units. Sol Energy 122:382–395
Liu M, Saman W, Bruno F (2012) Review on storage materials and thermal performance enhancement techniques for high temperature phase change thermal storage systems. Renew Sustain Energy Rev 16(4):2118–2132
Liu L, Su D, Tang Y, Fang G (2016) Thermal conductivity enhancement of phase change materials for thermal energy storage: a review. Renew Sustain Energy Rev 62:305–317
Lin Y, Jia Y, Alva G, Fang G (2018) Review on thermal conductivity enhancement, thermal properties and applications of phase change materials in thermal energy storage. Renew Sustain Energy Rev 82(3):2730–2742
Mousa MM, Bayomy AM, Saghir MZ (2020) Experimental and numerical study on energy piles with phase change materials. Energies 13:4699
Memon SA (2014) Phase change materials integrated in building walls: a state of the art review. Renew Sustain Energy Rev 31:870–906
Menter FR (1994) Two-equation eddy-viscosity turbulence models for engineering applications. AIAA J 32:1598–1605
Nithyanandam K, Pitchumani R (2013) Computational studies on a latent thermal energy storage system with integral heat pipes for concentrating solar power. Appl Energy 103:400–415
Oya T, Nomura T, Okinaka N, Akiyama T (2012) Phase change composite based on porous nickel and erythritol. Appl Therm Eng 40:373–377
Peng S, Fuchs A, Wirtz RA (2004) Polymeric phase change composites for thermal energy storage. J Appl Polym Sci 3(3):1240–1251

Pizzolato A, Sharma A, Maute K, Sciacovelli A, Verda V (2017) Design of effective fins for fast PCM melting and solidification in shell-and-tube latent heat thermal energy storage through topology optimization. Appl Energy 208:210–227

Porto TN, Delgado JMPQ, Guimarães AS, Magalhães HLF, Moreira G, Correia BB, Andrade TF, Lima AGB (2020) Phase change material melting process in a thermal energy storage system for applications in buildings. Energies 13:3254. https://doi.org/10.3390/en13123254

Shabgard H, Zhu W, Faghri A (2019) Integral solution of two-region solid–liquid phase change in annular geometries and application to phase change materials–air heat exchangers. Energies 12:4474

Sharma S, Buddhi D, Sawhney R, Sharma A (2000) Design, development and performance evaluation of a latent heat storage unit for evening cooking in a solar cooker. Energy Conv Manage 41(14):1497–1508

Sharma A, Chen C (2009) Solar water heating system with phase change materials. Int Rev Chem 1:297–307

Sharma A, Tyagi VV, Chen CR, Buddhi D (2009) Review on thermal energy storage with phase change materials and applications. Renew Sustain Energy Rev 13:318–345

Sharma RK, Ganesan P, Tyagi VV, Metselaar HSC, Sandaran SC (2015) Developments in organic solid-liquid phase change materials and their applications in thermal energy storage. Energy Conv Manag 95:193–228

Su W, Darkwa J, Kokogiannakis G (2015) Review of solid-liquid phase change materials and their encapsulation technologies. Renew Sustain Energy Rev 48:373–391

Tristão P (2014) Encapsulation of phase change materials (PCM) for energy storage applications. Master's Dissertation, Renewable Energies and Energy Efficiency, Polytechnic Institute of Bragança, Bragança, Portugal (in Portuguese)

Vaz Sá A, Azenha M, Guimarães AS, Delgado JMPQ (2020) FEM applied to building physics: Modeling solar radiation and heat transfer of PCM enhanced test cell. Energies 13:2200

Voller VR, Prakash C (1987) A fixed grid numerical modelling methodology for convection-diffusion mushy region phase-change problems. Int J Heat Mass Transf 30:1709–1719

Wahid MA, Hosseini SE, Hussen HM, Akeiber HJ, Saud SN, Mohammad AT (2017) An overview of phase change materials for construction architecture thermal management in hot and dry climate region. Appl Therm Eng 112:1240–1259

Waqas A, Kumar S (2013) Phase change material (Pcm)-based solar air heating system for residential space heating in winter. Int J Green Energy 10:402–426

Wilcox DC (1998) Turbulence modeling for CFD. DCW Industries: La Canada, CA

Xu Y, He YL, Li YQ, Song HJ (2016) Exergy analysis and optimization of charging-discharging processes of latent heat thermal energy storage system with three phase change materials. Sol Energy 123:206–216

Zalba B, Marín JM, Cabeza LF, Mehling H (2003) Review on thermal energy storage with phase change: Materials, Heat Transfer Analysis and Applications. Appl Thermal Eng 23:251–283

Zhao CY, Zhang GH (2011) Review on microencapsulated phase change materials (MEPCMs): fabrication, characterization and applications. Renew Sustain Energy Rev 15:3813–3832

Zhang P, Xiao X, Ma ZW (2016) A review of the composite phase change materials: Fabrication, characterization, mathematical modeling and application to performance enhancement. Appl Energy 165:472–510

Zhou D, Zhao CY, Tian Y (2012) Review on thermal energy storage with phase change materials (PCMs) in building applications. Appl Energy 92:593–605

Influence of the Coating System on the Acoustic Performance of Buildings

A. C. Azevedo, P. Freitas Gois, A. J. Costa e Silva, J. M. P. Q. Delgado, Marconi Barbosa, E. G. Remígio, and O. J. da Silva

Abstract Regardless of the construction processes or employed materials, the NBR 15575 (2013) defines objectives, requirements and methods for assessing the behaviour in use performance of accommodations. Therefore, the influence of the coating layers on the performance of construction systems is an essential parameter to assess the use of materials and elements of the systems, in order to make the project feasible, not only economically, but also technically. Through numerical simulations based on a defined reference model for the study, the present work study the influence of different layers of floor, roof and, internal and external wall systems, on the acoustic performance. Finally, analysing the results, for the reference model used, identified the materials and elements with the greatest influence on acoustic performance of the façade—function especially of the external windows; acoustic

A. C. Azevedo (✉) · J. M. P. Q. Delgado
Civil Engineering Department, CONSTRUCT-LFC, Universidade Do Porto, Rua Dr. Roberto Frias, 4200-465 Porto, Portugal
e-mail: antonio.costaazevedo@fe.up.pt

J. M. P. Q. Delgado
e-mail: jdelgado@fe.up.pt

P. F. Gois · A. J. C. Silva
Civil Engineering Department, Universidade Católica de Pernambuco, Recife, Brazil
e-mail: pedrodefg@gmail.com

A. J. C. Silva
e-mail: angelo@tecomat.com.br

M. Barbosa · E. G. Remígio · O. J. da Silva
Rua Serra da Canastra, 4TECOMAT Engenharia, Cordeiro, Recife, Pernambuco 39150640-310, Brazil
e-mail: mendonca.marconi@gmail.com

E. G. Remígio
e-mail: egremigio@gmail.com

O. J. da Silva
e-mail: otaviojsjunior@gmail.com

J. M. P. Q. Delgado (ed.), *Efficient and Suitable Construction*, Building Pathology and Rehabilitation 17, https://doi.org/10.1007/978-3-030-62829-1_2

performance of the floor system—mainly influenced by the thickness of the structural element and the use of ceiling and acoustic blanket; acoustic performance of internal walls—typology of the structured element of the wall and thickness.

Keywords Building performance · Acoustic performance · Coatings · Numerical simulations

1 Introduction

The connection between man and construction, especially housing construction, is remote and intertwined with the evolutionary history of humanity and society itself. The Code of Hammurabi, a set of laws created in Mesopotamia, around the eighteenth century BC, in the first Babylonian dynasty, was already about construction and its responsibilities. The aforementioned code of law, known as "an eye for an eye, a tooth for a tooth", in its set of rules already provided that if a builder erects a house, and the house collapses and kills the resident, the builder will be immolated—sacrificed—killed as revenge.

Taking into account that the Code of Hammurabi does not prescribe how the house should be built, the types of materials that should be used, the width, size and parts of the construction or the method of construction, defining only the final result that must be achieved: the construction cannot collapse and kill someone. The user behaviour of the construction is clearly defined, possibly, by the Hammurabi code. This code was the first standard of performance of housing buildings in the world, already establishing requirements, in this case for structural security. The establishment of functional requirements for buildings and their parts stems from the obvious premise that buildings, being indispensable to the life and activity of man, must have characteristics that correspond to and meet human needs. In other words, the establishment of functional requirements for buildings is a performance prescription (Gomes 2015).

The word performance is widely used throughout society and has quite broad meaning. It is used for hardware evaluation, professional analysis and business and sports, for example. It is common to use performance to compare professionals and equipment and, in general, a desirable standard is defined, often informally, for comparison with the performance delivered. A more modern view of performance began to be structured in the twentieth century, where studies were proposed by the National Bureau of Sciences (NBS) during the 1920s. In the 1930s and 1940s, the first performance standards were developed and the English expression performance requirements emerged (Lorensi 2015).

After the Second World War and the consequent need to build large-scale buildings in the reconstruction movement, especially in Europe, the application of innovative construction technologies and systems at the time caused the incidence of high cases of pathological manifestations, generating high economic and social burdens. Given this scenario, the need for a more careful analysis of the performance of the construction systems used proved to be very relevant. From the end of the 1960s, the United States of America and some European countries devoted themselves to deepening their studies and striving to solidify the application of the concept of performance to buildings, the book "*Savoir batir: habitabilite, durabilite, economie des batiments*", by Gerard Blachére recognized as an important publication on the theme, which conceptualizes the performance of buildings as the behaviour in use during a given useful life (Souza et al. 2018).

ISO 6241 (1984) "Performance standards in building—Principles for their preparation and factors to be considered" is an important regulatory framework for building performance. ISO 6241 (1984) establishes general principles for the development of performance standards in civil construction expressed in functional requirements of users, linking the performance of buildings and user requirements. The objective of this Standard was to assist ISO signatory countries in the elaboration of Performance Standards, and to serve as a guide for the selection of requirements that can be applied in each case when talking about building performance (Borges 2008). Table 1 presents the requirements of users considered in ISO 6241.

ISO 6241 (1984) is still a valid and important reference for defining the performance requirements of buildings, and perhaps its main gap in relation to contemporary requirements is sustainability, considering that at the time the theme did not have the current relevance.

Like what happened in Europe in the post-war period, in the 1970s and 1980s in Brazil, the construction of large-scale buildings induced the use of new techniques and constructive technologies. During this period, known as the "Brazilian miracle", productivity was prioritized, without clear technical criteria for evaluating the innovations adopted to enable this productivity.

Also in the 1970s and 1980s, studies on the performance of buildings were developed by IPT—Institute of Technological Research, at the request of BNH—*Banco Nacional de Habitação*, studies on the performance of buildings, considered as the first studies on the subject in Brazil.

However, the first publications on the subject in Brazil report inadequate uses of innovative construction techniques and systems related to users' functional requirements, and related to the conditions of exposures exposed to buildings. According to Melo (2007), most of the innovative solutions implemented in the country, mainly in the construction of sets financed by the extinct BNH, had components and construction systems introduced without them having an adequate technical evaluation so that they could thus predict their behaviour during their useful life. Thus, in these cases, there was an evaluation of post-occupation performance, where users served as "guinea pigs", and they were penalized with pathological problems and maintenance and replacement costs that were a consequence of the use of new poorly developed products and without adequate technical evaluation.

In resume, the same process of development of technological innovations of post-war components and construction systems in Europe occurred in Brazil, without proper performance analysis and with a high incidence of pathological manifestations causing an economic and social burden for all actors in the construction chain.

Table 1 Users requirements

Category	Examples
1. Stability requirements	Mechanical resistance to static and dynamic actions, both individually and in combination. Resistance to impacts, intentional or unintentional abusive actions, accidental actions, cyclical effects
2. Fire safety requirements	Risks of fire and fire diffusion, respectively. Psychological effects of smoke and heat Alarm triggers time (detection and alarm systems). Building evacuation time (exit routes). Survival time (compartmentalization of fire)
3. Safety risks in use	Safety related to aggressive agents (protection against explosions, burns, sharp points and edges, moving mechanisms, electrical discharges, radioactivity, contact or inhalation of poisonous substances, infection). Safety during movement and movement (limitation of slipping on floors, unobstructed roads, handrails, etc.). Security against the improper entry of persons and/or animals
4. Sealing requirements	Water sealing (rain, subsoil, drinking water, wastewater, etc.). Air and gas seal. Dust and snow seal
5. Thermal and moisture requirements	Control of air temperature, thermal radiation, air speed and relative humidity (limitation of variation in time and space, response of controls). Condensation controls
6. Air purity requirements	Ventilation. Odour control
7. Acoustic requirements	Control of internal and external noise (continuous and/or intermittent). Sound intelligibility. Reverberation time
8. Visual requirements	Natural and artificial lighting (necessary lighting, stability, luminous contrast and very strong light protection). Sunlight (heat stroke). Possibility of darkness. Aspects of spaces and surfaces (colour, texture, regularity, levelling, verticality, horizontality, etc.). Visual contact, internally and with the outside world (threads and barriers related to privacy, protection against optical distortion)
9. Tactile requirements	Properties of surfaces, roughness, dryness, heat, elasticity. Protection against static electricity discharges

(continued)

Table 1 (continued)

Category	Examples
10. Dynamic requirements	Limitation of vibrations and accelerations of the whole set (transient and continuous). Convenience of pedestrians in areas exposed to the wind. Ease of movement (slope of ramps, arrangement of stairs steps). Room for manoeuvre (handling of doors, windows, control over equipment, etc.)
11. Hygiene requirements	Installation for care and hygiene of the human body. Water supply. Cleaning conditions. Release of wastewater, materials served and smoke. Limitation of contaminant emission
12. Requirements for the convenience of spaces intended for specific uses	Quantity, size, geometry, subdivision, and interrelation of spaces. Services and equipment. Conditions (capacity) of furniture and flexibility
13. Durability requirements	Conservation (permanence) of performance in relation to the necessary service life subject to regular maintenance
14. Economic requirements	Maintenance, operational and capital costs. Demolition costs

As an example of this process of technological innovations without a more detailed performance analysis, the cases of the "coffin buildings" built in the Metropolitan Region of Recife are mentioned. Melo (2007) mentions that since 1977, landslides have been recorded in 12 buildings in Recife and 50 buildings were still banned in Olinda. The study also recalled that estimates from ITEP—*Instituto Tecnológico de Pernambuco* state that at least one in ten thousand buildings in the state can collapse. A preoccupant number of buildings could collapse, in view of the consequence of this performance failure.

According to Lorensi (2015), as a result of the events, and with the focus on promoting the quality of works and leveraging the civil construction sector, CEF—*Caixa Econômica Federal* commissioned to IPT a study on the subject, which ran from 1981 to 1997. One of the pioneering works on the subject was published in the book "Building Technology", launched by IPT in 1988, with emphasis on evaluating the performance of construction systems for housing buildings. During this period, according to Manso and Filho (2007), the IPT developed, with the support of the Studies and Projects Financier (FINEP), the work entitled "Minimum Performance Standards", completed in 1995.

In 1997, *Caixa Econômica Federal* hired the IPT to review the work done in 1981, and other studies were done in the same way as that prepared in 1999 by the Brazilian Institute of Technology and Construction Quality. Considering the existence of several independently developed references, Caixa Econômica Federal and the technical environment identified the need to harmonize them, transforming them into technical standards that would further facilitate the evaluation process. For the elaboration of these Standards, Caixa Econômica Federal, with the support of Finep, financed the research project "Technical Standards for the Evaluation of Innovative Construction Systems for Housing", in 2000 (Borges 2008).

In 2000, a Study Committee and working groups were created with the objective of coordinating the discussion on the performance of buildings in the technical environment, seeking consensus for the transformation of the product into A Brazilian Standard, within the scope of ABNT. The coordinator elected to the Study Commission in 2000 was Engineer Ércio Thomaz, ipt. In a second moment, in 2004, a new coordinator was elected to the Commission for The Study of the Performance Standard Project, Carlos Alberto de Moraes Borges, who remained in office until the publication of the first version of the said Standard, on May 12, 2008, and its enforceability is expected in 2010.

In 2010, with the proximity of the expected date for effective enforceability of the Performance Standard, the construction chain understood as necessary the expansion of the discussion and a longer adaptation period so that actors in the construction chain could adapt to the requirements of the Standard. Thus, in practice, the ABNT Version NBR 15575 (2008) did not come into force and had its enforceability extended to 2012. Also in 2010 was elected a new coordinator for the Study Committee, Fábio Villas Boas, who conducted a new process of discussion and revision of the text of the Standard. The new text was put into public consultation in 2012 and was published and entered into force in 2013. This version of the Standard remains in force to date, however a new study committee is open, still under the coordination of Fábio Villas Boas, conducting discussions with the aim of revising the Standard and publishing a new version in the year 2020.

1.1 Performance Standard—NBR 15575 (2013)

The popularly known Performance Standard is a technical standard of ABNT—Brazilian Association of Technical Standards, NBR 15575—Housing Buildings—Performance. It is noteworthy that NBR 15575 is divided into 6 parts, the first of which deals with the general requirements of the building and the following deals with the requirements of its parts, represented by systems. This division makes clear the concept of analysis of the overall analysis of the building and the parts that compose it. So the six parts of NBR 15575 are:

- NBR 15575-1 (2013): General requirements (Brazilian Association of Technical Standards. NBR 15575-1: Housing buildings—Performance. Part 1: General Requirements. Rio de Janeiro, 2013).
- NBR 15575-2 (2013): Requirements for structural systems (Brazilian Association of Technical Standards. NBR 15575-2: Housing buildings—Performance. Part 2: Requirements for structural systems. Rio de Janeiro, 2013).

- NBR 15575-3 (2013): Requirements for flooring systems (Brazilian Association of Technical Standards. NBR 15575-3: Housing buildings—Performance. Part 3: Requirements for flooring systems. Rio de Janeiro, 2013).
- NBR 15575-4 (2013): Requirements for internal and external wall systems—IWS and EWS (Brazilian Association of Technical Standards. NBR 15575-4: Housing buildings—Performance. Part 4: Requirements for internal and external wall systems. Rio de Janeiro, 2013).
- NBR 15575-5 (2013): Requirements for roofing systems—RS (Brazilian Association of Technical Standards. NBR 15575-5: Housing buildings—Performance. Part 5: Requirements for roofing systems. Rio de Janeiro, 2013).
- NBR 15575-6 (2013): Requirements for hydro sanitary systems (Brazilian Association of Technical Standards. NBR 15575-6: Housing buildings—Performance. Part 6: Requirements for hydro sanitary systems. Rio de Janeiro, 2013).

It can be observed by dividing the parts of NBR 15575 that the performance analysis is specified for the building and its systems. Therefore, understanding the concept of system is very important and the standard itself defines system as "most functional part of the building. Set of elements and components intended to meet a macro-function that defines it."

Figure 1 exemplifies a floor system, where it can be intuitively identified that the final performance of the system will depend on the characteristic of the components and elements that compose it.

According to the definition of NBR 15575-1 (2013), performance is the "behaviour in use of a building and its systems". This definition makes clear the concept of scope for all housing buildings, regardless of the construction systems, elements and components used, because the object of the Standard is behaviour in use of the building and its parts. This bias is different from most ABNT technical standards related to civil construction, which focuses on the prescription of methods of sizing and execution of specific components, elements and construction systems.

The performance definition of NBR 15575 also explains the intimate relationship with the requirements of users and the exposure conditions to which the building is exposed, and the main objective of NBR 15575 is to establish requirements and performance criteria applicable to housing buildings, as an integrated whole, as well as to be evaluated in insulation for one or more specific systems, represented by each part of the Standard.

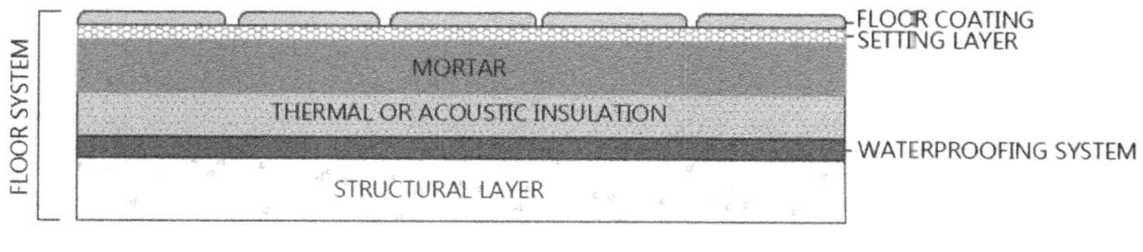

Fig. 1 Generic example of a floor system and its elements *Source* NBR 15575-3, (2013)

Performance requirements, according to the Standard itself, represent the conditions that qualitatively express the attributes that the housing building and its systems must have, so that they can meet the needs of the user. Using acoustic façade performance as an example, the user's requirement is the acoustic insulation of noise stemming from the exterior of the building. The requirement does not express values, being naturally qualitative.

The definition of the performance requirements of NBR 15575 followed the conceptual line of ISO 6147 considering the understanding that the requirements set forth in the Standard represent the requirements of users. The requirements set out below are grouped by category: Safety, Habitability and Sustainability.

- Safety
  - Structural safety
  - Fire safety
  - Safety in use and operation
- Habitability
  - Watertightness
  - Thermal performance
  - Acoustic performance
  - Lighting performance
  - Health, hygiene and air quality
  - Functionality and accessibility
  - Tactile and anthropodynamic comfort
- Sustainability
  - Durability and maintainability
  - Environmental impact

Table 2 shows the number of specific requirements for each requirement and performance category.

Given the need to present values, the standard establishes performance criteria for each requirement. By definition, criteria are quantitative specifications of performance requirements, expressed in terms of measurable quantities, so that they can be objectively determined. Following the example of acoustic performance of facades, the criterion for the requirement of acoustic insulation of external noises established by the standard is 25 decibels, considering noise class II, which represents the exposure condition.

That is, the criterion expresses a value for the performance requirement, which makes it tangible, measurable, and for the measurement of these values are established, for each requirement and performance criterion, evaluation methods. The objective of the evaluation methods is to ensure the uniformity and representativeness of the measurements and, consequently, of the analyses. That is, regardless of who is responsible for measuring the acoustic insulation of the façade, for example,

Table 2 Number of specific performance requirements contained in NBR 15575 separated by requirement and category

Category	Requirements	No. of specific requirements	
Safety	Structural safety	29	68
	Fire safety	20	
	Safety in use and operation	19	
Habitability	Water tightness	16	60
	Thermal performance	6	
	Acoustic performance	12	
	Lighting performance	3	
	Tactile and anthropodynamic comfort	4	
	Health, hygiene and air quality	9	
	Functionality and accessibility	10	
Sustainability	Durability and maintainability	19	21
	Environmental impact	2	
Total		149	

the result should be the same, because the method used is standardized and established in NBR 15575 itself.

In resume, the methodology of performance analysis necessarily passes through the tripod requirement, criterion and evaluation method. It is noteworthy that the requirement is an expression of the users' requirements and the exposure conditions to which the building and its parts are exposed.

Figure 2 presents schematically the methodology of performance evaluation.

Another important point defined in NBR 15575, which brings an important legal relevance to the norm, is the attribution of tasks and responsibilities of the actors

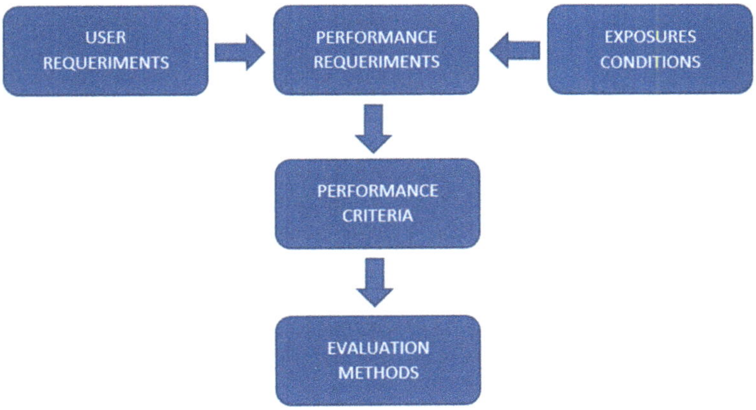

Fig. 2 Performance evaluation methodology

involved in the design, construction and use of the building. NBR 15575 classifies and assigns the following responsibilities to the actors:

- Supplier of materials, components and/or systems: It is up to the system supplier to characterize the performance, according to NBR 15575. Suppliers of components and elements shall characterize them in accordance with their applicable prescriptive standards and, if they do not exist, provide supporting results of performance based on NBR 15575 and specific international or foreign standards.
- Designer: The designer has the role of specifying materials, products and processes that meet the performance established based on prescriptive standards and the performance declared by the manufacturers of the products to be used in the project.
- Builder and developer: It is up to the developer to provide specific technical studies to identify foreseeable risks at the time of design. The builder or developer is responsible for the preparation of the manual of use and operation of the building to be delivered to the owner of the housing unit and the liquidator, in the case of the manual of use, operation and maintenance of the common areas.
- User: The user is responsible for performing maintenance in accordance with the provisions of NBR 5674 (2012) and the manual of use, operation and maintenance of the building.

Presenting responsibilities of the actors emphasizes the need for knowledge about the characteristics of the components and elements, including the coating layers, which make up the systems, mainly by designers and suppliers of components, elements and construction systems. "The quality of the products associated with a permanent increase in efficiency will certainly be attributes that will differentiate and privilege the companies that act with these criteria in the various sectors of the construction production chain, whose initial link, or starting point, is the design (architectural and engineering), responsible for defining the building in its fundamental characteristics (shape/design, construction system, component subsystems etc.)".

1.2 Acoustic Performance of Housing Buildings

"Housing building must have adequate sound insulation of external fences, with regard to air noise stemming from the exterior of the housing building, and adequate soundproofing between common and private areas of different autonomous units" (NBR 15575-1, 2013). The definition of NBR 15575-1 (2013) indicates the need for insulation from noise generated outside the housing unit, whether generated by a neighbouring unit or common area of the building itself or by sources outside the building. Analysing the standard we can divide the acoustic performance requirement into four groups:

- Insulation of external walls to air noise;

- Insulation of internal walls to overhead noise;
- Insulation of floor systems from overhead noise;
- Insulation of floor systems to impact noises.

Before detailing the criteria for each requirement listed above, it is important to understand the evaluation methods recommended in the Performance Standard:

- Precision method performed in the laboratory: Determines the sound insulation of components and building elements (wall, window, door and others), providing calculation reference values for projects. The result obtained in the laboratory test is the weighted sound reduction index, represented by the Rw symbol, as shown in Table 3.
- Engineering method carried out in-field: Determines in-field, rigorously, the overall sound insulation of the walls, directly characterizing the acoustic behaviour of the system. The engineering method is, in fact, the valid method for proving acoustic performance, because it considers the overall behaviour of the building in its conditions of use. In-field measurements are represented by three symbols, referring to the measured parameter and the respective type of test performed, as shown in Table 3.
- Simplified in-field method: This method allows an estimate of the overall noise insulation to airborne noise of the wall systems in situations where the necessary equipment for the execution of the test by the engineering method is not available. This method has no normative value, that is, it does not serve as proof of performance.

After the requirements and methods of acoustic performance evaluation of the Performance Standard, it is important to present the criteria, the values for quantification of requirements. With regard to the isolation of air noise from external walls, a decisive factor for defining the evaluation criterion is surrounding noise class, which represents the exposure condition of the building. Considering that a housing unit

Table 3 Acoustic parameters for performance verification

Symbol	Description	Standard	Application
R_w	Weighted sound reduction index	ISO 10140-2 ISO 717-1	Components in the laboratory
$D_{nT,w}$	Standardized weighted level difference	ISO 16,283-1 ISO 717-1	Internal vertical and horizontal seals, in the laboratory (walls, floors, etc.)
$D_{2m,nT,w}$	Standardized difference of weighted level 2 m away from the façade	ISO 16283-3 ISO 717-1	Facades in buildings Facades and roofs in storey houses
$L'_{nT,w}$	Weighted standard impact sound pressure level	ISO 16283-2 ISO 717-2	Floor system

Note: R_w–weighted sound reduction index, $D_{nT,w}$–weighted standardized level difference, $D_{2m,nT,w}$–weighted standardized level difference at 2 m, $L'_{nT,w}$–weighted standardized impact pressure level

Table 4 Minimum values of the standardized difference of weighted level, $D_{2m,nT,w}$, of the bedroom external wall (NBR 15575-4, 2013)

Noise class	Housing location	$D_{2m,nT,w}$ (dB)
I	Housing located far from sources of intense noise	≥20
II	Housing located in areas subject to non-fit noise situations in Class I or III	≥25
III	Housing subject to intense noise from transports and others, since that it complies with the legislation	≥30

Notes: (1) For external walls of rooms, kitchens, laundries and bathrooms, there are no specific requirements, (2) In regions of airports, stadiums, sporting event venues, highways and railways there is a need of specific studies.

exposed to a higher noise level will have, consequently, to have a greater isolation, Table 4 presents the criteria for insulation of external walls (façades), according to the surrounding noise class.

Table 4 shows that the insulation criterion to external noise is applicable only to the bedrooms and that the definition of the location and, consequently, of the noise class, is quite subjective.

Taking into account the subjectivity in the definition of the surrounding noise class and the buildings located near airports, stadiums, highways, railways and other environments in which the Performance Standard recommends specific studies for the classification of surrounding noise, *Proacústica*—Brazilian Association for Acoustic Quality—has launched a Manual with the objective of bringing clearer information on the subject. Among other information, the Proacoustic Manual for noise class of housing buildings (PROACÚSTICA 2017) establishes an objective criterion for defining the surrounding noise class, based on the equivalent sound pressure level, $L_{Aeq,T}$, incident on the facades, as presented in Table 5, adapted from the publication and presented below.

It is important to note that the surrounding noise is dynamic, and may vary with the urban development of the city, either by the implementation of new roads, traffic changes or mainly by real estate expansion, as in the case of the implementation of housing estates in previously poorly inhabited areas, as proven by (Remigio et al. 2019). Regarding the insulation to aerial noise scans of internal walls, the Performance Standard establishes criteria for twinning walls, which divide distinct housing

Table 5 Equivalent sound pressure levels, $L_{Aeq,T}$, incidents on the facades of buildings for each noise class (PROACÚSTICA 2017)

Noise class	Equivalent sound pressure levels $L_{Aeq,T}$(dB)	$D_{2m,nT,w}$ (dB) (In-field testing engineering method)
I	< 60	≥20
II	61–65	≥25
III	66–70	≥30

Table 6 Minimum values of the standardized weighted level difference, $D_{nT,w}$, between environments (NBR 15575-4 2013)

Element	$D_{nT,w}$ (dB)
Wall between autonomous housing units (twinning wall), in situations where there is no bedroom environment	≥40
Wall between autonomous housing units (twinning wall), in case at least one of the environments is dormitory	≥45
Blind wall of bedrooms between a housing unit and common areas of eventual transit, such as corridors and staircase on the floors	≥40
Blind wall of rooms and kitchens between a housing unit and common areas of eventual transit, such as corridors and staircase on the floors	≥30
Blind wall between a housing unit and common areas of permanence of people, leisure activities and sports activities, such as home theatre, gyms, ballroom, games room, bathrooms and locker rooms collective, kitchens and collective laundries	≥45
Set of walls and doors of distinct units separated by the hall ($D_{nT,w}$ obtained between the units)	≥40

units, blind walls that divide housing units and common areas, and set of walls and doors of distinct units separated by the hall, as presented in Table 6.

It can be observed that the criteria are more rigorous for bedrooms environments, rest environments, where naturally, due to the very nature of use, the user needs a lower level of sound pressure and, consequently, a greater sound insulation.

In relation to floor systems, according to the requirements presented above, in addition to air noise insulation, criteria are prescribed for noise insulation of impacts set out in Tables 7 and 8.

Table 7 shows that, in addition to the stricter criterion for the acoustic insulation of the bedrooms, equivalent to the criteria for walls.

Greater air noise insulation is also required for floor systems that separate autonomous housing units from areas of collective use of prolonged permanence, which is explained by the conditions of use of these environments, which naturally has a higher noise generation potential.

Table 7 Weighted level standardized difference criteria, $D_{nT,w}$ (NBR 15575-4 2013)

Element	$D_{nT,w}$ (dB)
Floor system between autonomous housing units, in case at least one of the environments is a bedroom	≥45
Floor system separating autonomous units from common areas of eventual transit, such as corridors and staircase on the floors, as well as on different floors. Floor system between autonomous housing units, in situations where there is no bedroom	≥40
Floor system separating autonomous housing units from common areas for collective use, for leisure and sports activities, such as home theatre, gyms, ballroom, games room, bathrooms and collective changing rooms, kitchens and collective laundries	≥45

Table 8 Criterion and sound pressure level of weighted standard impact, $L'_{nT,w}$ (NBR 15575-4 2013)

Element	$L'_{nT,w}$ (dB)
Floor system separating autonomous housing units positioned on different floors	≤80
Floor system of public use areas (leisure and sports activities, such as home theatre, gyms, ballroom, games room, bathrooms and collective changing rooms, kitchens and collective laundries) on autonomous housing units	≤55

It is important to highlight that for the analysis of insulation to impact noise, differently from what occurs for air noise, the difference in sound pressure level between emission and reception environments is not measured. The method for measuring impact noise insulation is the direct measurement of the standard impact sound pressure level weighted, and the noise is generated by standard equipment defined at ISO 16283-2. That is, the higher the sound pressure level of impact measured, the worse the performance of the floor system.

Following the same understanding that the areas of collective use of prolonged permanence have a greater potential for generating noise exposed in the analysis of air noise insulation, it can be observed in Table 8 that the criterion for floor systems of common areas on housing units is also more rigid. However, in the case of noise insulation of impacts, the numerical difference of the value is much greater, of 25 dB.

It is important to highlight that the criterion of sound insulation refers to the wall system, and all the elements and components that compose it, such as, type of block in the case of masonry, frames, structural element, layers of coatings, have influence on the final acoustic insulation. That is, it is important to know the contribution of each layer to size the wall system to meet the criteria prescribed in the Performance Standard.

Therefore, for systems analysis it is important to understand the main characteristics of materials that interfere in the acoustic insulation of wall systems.

The variation of sound pressure by which the building elements are submitted, causes them to vibrate and this vibration is controlled mainly by the surface mass. Souza et al. (2018) highlight that the mass of the material influences the efficiency of the acoustic insulation of the elements, however, the importance of the mass depends on the sound frequency, since for low frequency sounds, the mass increase is less efficient than for high frequency sounds. In this sense, it can be said that the study of sound insulation has a close relationship with the law of the masses.

The law of the masses, presented in the previous paragraph, has an excellent correlation with the insulation considering the air noises; however, it does not have satisfactory application for noise generated by impacts, which are transmitted mainly by the vibration of the elements of the wall system itself, as illustrated in Fig. 3.

Several studies have been done to seek better methods of isolating impact noise from the floor, and the use of floating floors have been the most effective in this regard (Souza et al. 2018).

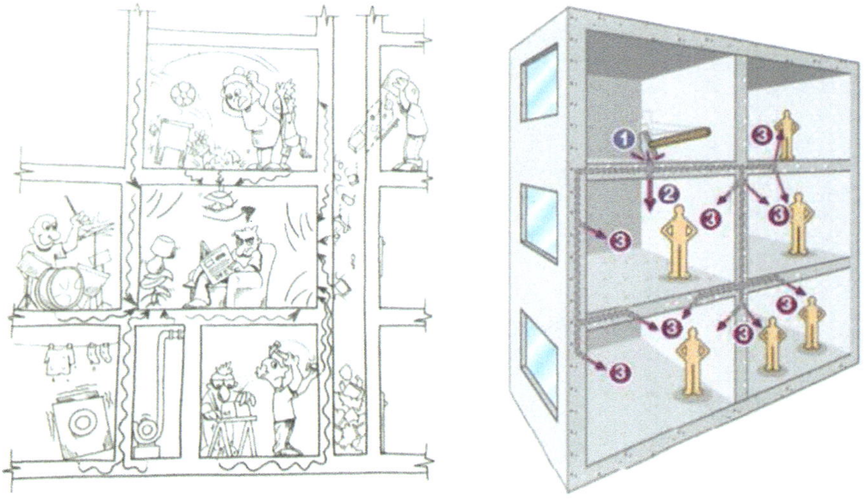

Fig. 3 Illustrations of transmission of noise generated by impacts

2 Methodology

The methodology adopted in this work initially consisted of the definition of a reference model for the numerical simulations of acoustic performance. Such a reference model has defined geometry, orientation and vertical and horizontal sealing systems.

The study was based on the variation of the layers of the vertical and horizontal sealing systems, mainly cladding layers, and comparative analysis of performance in relation to the reference model.

As previously described, the present work contemplates the evaluation of the influence of the coating systems on the natural thermal, acoustic and lighting performance of buildings. The experimental study was carried out from numerical simulations (as prescribed in the different specific regulations) carried out on the 3 referred requirements comparing the use of different compositions of the coating system, especially type of material, colour and thickness.

In order to compare the influence of the coating systems on the final performance relative to the influence of other subsystems, elements and components of the wall systems, variations in structural elements of the walls, windows, thickness of the structural element of the floor system were considered, type of roofing system and thermal and acoustic insulation.

In order to allow a comparative evaluation, a reference model environment was adopted, with typical dimensions of a room with the following internal dimensions: width equal to 2.60 m; length equal to 3.20 m; and ceiling height equal to 2.60 m.

For the analysis of thermal performance, the computational simulation method prescribed in NBR 15575-1 (2013) was used. For acoustic performance analysis, the computational simulation method prescribed at ISO 12354-1 (2017), ISO 12354-2

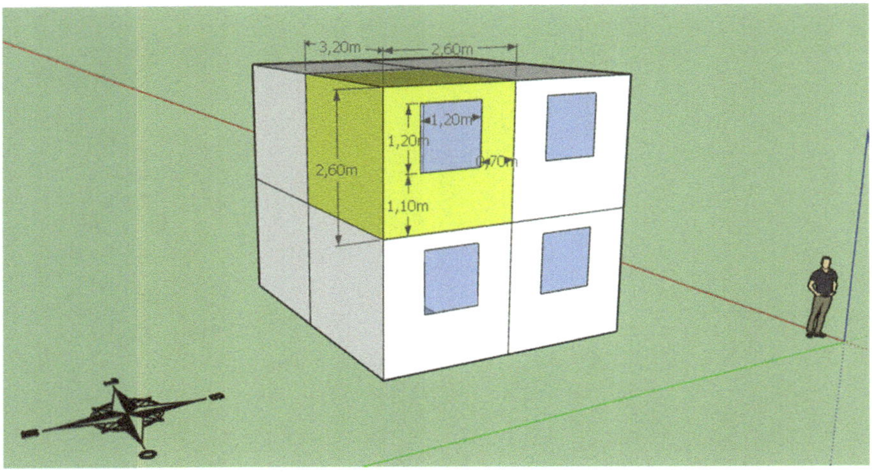

Fig. 4 Reference model used in numerical simulations

(2017) and ISO 12354-3 (2017) were used. For natural lighting performance analysis, the computational simulation method prescribed in NBR 15575-1 (2013) was used. For the location of the building and, consequently, the temperature and lighting data, the city of Recife—PE, contained in the bioclimatic region 8, was considered. The orientation of the model followed the critical orientation regarding thermal performance in the summer, following the recommendation of item 11.5.1 of NBR 15575-1 (2013): "bedroom or living room window facing south and the other exposed wall facing north".

The reference model used in the study did not consider the existence of vegetation or buildings in the surroundings, although the influence of shading is known and relevant in the performance of the actual building. Corroborating this statement, in a research conducted in the northern hemisphere (Chan 2012) indicates that the effect caused by adjacent apartments also reduces the gain of solar heat in cold season, resulting in an increased need for energy for heating. It was considered in the façade of the model the use of sliding frames with two movable sheets with dimensions 1.20 × 1.20 m, typical of housing buildings of economic standard. Figure 4 presents a sketch of the reference model used for numerical simulations. The environment filled in yellow was evaluated in the simulations.

2.1 *Façades*

For the external wall system (façade) of the reference model it was considered masonry of ceramic blocks of 8 holes horizontally (9 × 19 × 19 cm), internally

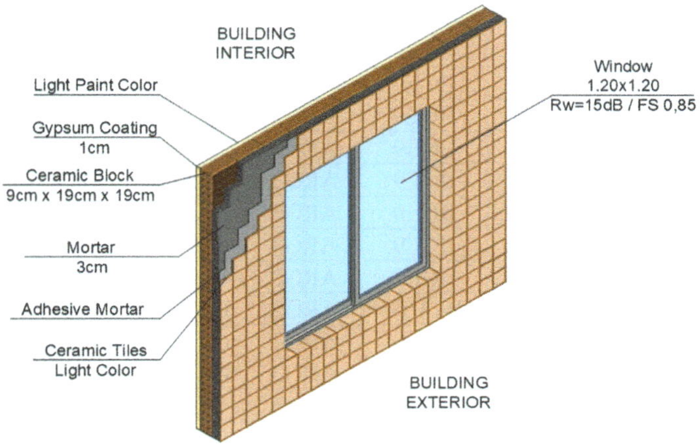

Fig. 5 Schematic detail of the external wall system of reference used in the study

coated with gypsum paste (thickness equal to 1 cm) and light color paint and externally coated with cementitious mortar (thickness equal to 3 cm), and light coloured ceramic board. For the windowsr, the reference model was considered the typology of running with two movable sheets, with sound reduction index equal to 15 dB (Rw = 15 dB) and colorless float glass of 4 mm (Fig. 5).

In addition to the external wall system defined for the reference model, the following variations were adopted for the elements/components with the objective of comparative analysis:

- 2 colours for the finishing of the internal coating;
- 4 variations in the internal coating, with changes in the type of material and thickness;
- 3 types of wall structuring elements;
- 2 thicknesses for external coating;
- 5 variations in the external finishing layer, with changes in the type of material and colours;
- 2 types external frames (windows), with variations in the Sound Reduction Index (Rw) and Solar Factor (FS) of the glasses (Table 9).

2.2 *Internal Wall Systems*

For the internal vertical wall system of the reference model, masonry of 8-hole ceramic blocks was considered horizontal 9 × 19 × 19 cm coated with gypsum paste (thickness $\varepsilon = 1$ cm) and light colour paint on both sides.

Figure 6 presents schematic detail of the internal vertical wall system considered in the reference model.

Table 9 External wall systems (EWS) considered in numerical simulations

EWS	Internal coating		Wall	External coating		Window
1	PINα0.3	GES1	BCE9	ARG3	CERα0.3	ESQRW15FS0.85
2	PINα0.5	GES1	BCE9	ARG3	CERα0.3	ESQRW15FS0.85
3	PINα0.7	GES1	BCE9	ARG3	CERα0.3	ESQRW15FS0.85
4	PINα0.3	GES3	BCE9	ARG3	CERα0.3	ESQRW15FS0.85
5	PINα0.3	ARG2	BCE9	ARG3	CERα0.3	ESQRW15FS0.85
6	PINα0.3	ARG4	BCE9	ARG3	CERα0.3	ESQRW15FS0.85
7	PINα0.3	ARG6	BCE9	ARG3	CERα0.3	ESQRW15FS0.85
8	PINα0.3	GES1	BCE14	ARG3	CERα0.3	ESQRW15FS0.85
9	PINα0.3	GES1	BCO14	ARG3	CERα0.3	ESQRW15FS0.85
10	PINα0.3	GES1	PCO10	ARG3	CERα0.3	ESQRW15FS0.85
11	PINα0.3	GES1	BCE9	ARG5	CERα0.3	ESQRW15FS0.85
12	PINα0.3	GES1	BCE9	ARG7	CERα0.3	ESQRW15FS0.85
13	PINα0.3	GES1	BCE9	ARG3	CERα0.5	ESQRW15FS0.85
14	PINα0.3	GES1	BCE9	ARG3	CERα0.7	ESQRW15FS0.85
15	PINα0.3	GES1	BCE9	ARG3	TEXα0.3	ESQRW15FS0.85
16	PINα0.3	GES1	BCE9	ARG3	TEXα0.5	ESQRW15FS0.85
17	PINα0.3	GES1	BCE9	ARG3	TEXα0.7	ESQRW15FS0.85
18	PINα0.3	GES1	BCE9	ARG3	CERα0.3	ESQRW19FS0.66
19	PINα0.3	GES1	BCE9	ARG3	CERα0.3	ESQRW23FS0.52

Legend: **PINα0.3 | PINα0.5 | PINα0.7**: Painting with light colour ($\alpha = 0.3$), medium ($\alpha = 0.5$) and dark ($\alpha = 0.7$), **GES1 | GES3**: Gypsum plaster with thicknesses of 1 cm and 3 cm, **ARG2 | ARG3 | ARG4 | ARG5 | ARG6 | ARG7**: Mortar with thicknesses from 2 to 7 cm, **BCE9**: Masonry of 8-hole ceramic blocks horizontally 9 × 19 × 19 cm, **BCE14**: Masonry of ceramic blocks with vertical hole 14 × 19 × 39cm, **BCO14**: Concrete block masonry with vertical hole 14 × 19 × 39cm, **PCO10**: Solid concrete wall 10 cm thick, **CERα0.3 | CERα0.5 | CERα0.7**: Ceramic boards with light, medium and dark colour, **TEXα0.3 | TEXα0.5 | TEXα0.7**: Acrylic texture with light, medium and dark colour, **ESQRW15FS0.85**: Window Rw = 15 dB and colourless Float glass 4 mm (Solar Factor = 0.85), **ESQRW19FS0.66**: Window Rw = 19 dB and green Float glass 4 mm (Solar Factor = 0.66), **ESQRW23FS0.52**: Window Rw = 23 dB and grey laminated glass 6 mm (Solar Factor = 0.52)

Notes: Line 1 represents the reference model and cells filled in yellow represent the layer of the wall system changed from the reference model

In addition to the internal wall system defined for the reference model, the following variations were adopted for the elements/components for comparative analysis:

- 2 colours for finishing the inner coating;
- 4 variations in the inner coating, with changes in the type of material and thickness;
- 3 types of wall structuring elements.

The reference model and the variations adopted were considered in the models with nine typologies of internal wall systems.

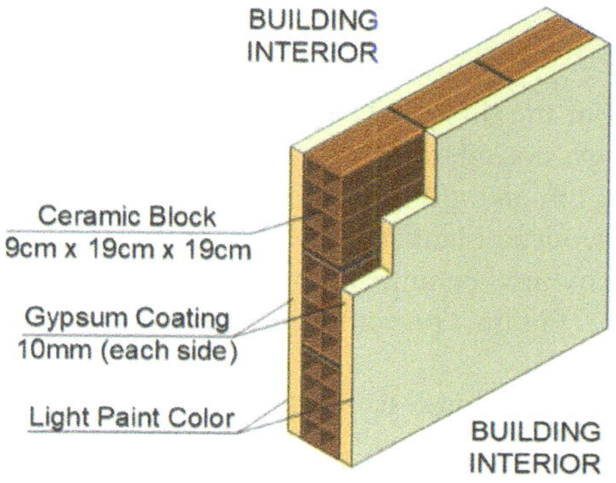

Fig. 6 Schematic detail of the reference internal wall system used in the study

Table 10 presents the relationship of internal wall systems, considering the variations previously pointed out.

The ceramic and concrete blocks with dimensions 14 × 19 × 39 cm considered in the simulations have vertical bore and dimensions according to NBR 15575-1 (2017) and NBR 6136 (2016), respectively.

This information is relevant because the soundproofing properties of perforated blocks are impaired compared to solid blocks and that different geometries can provide also different sound insulation (Fringuellino and Smith 1999).

Table 10 Internal wall systems (IWS) considered in numerical simulations

IWS	Coating 1		Wall	Coating 2	
1	PINα0.3	GES1	BCE9	GES1	PINα0.3
2	PINα0.5	GES1	BCE9	GES1	PINα0.5
3	PINα0.7	GES1	BCE9	GES1	PINα0.7
4	PINα0.3	GES3	BCE9	GES3	PINα0.3
5	PINα0.3	ARG2	BCE9	ARG2	PINα0.3
6	PINα0.3	ARG4	BCE9	ARG4	PINα0.3
7	PINα0.3	ARG6	BCE9	ARG6	PINα0.3
8	PINα0.3	GES1	BCE14	GES1	PINα0.3
9	PINα0.3	GES1	BCO14	GES1	PINα0.3
10	PINα0.3	GES1	PCO10	GES1	PINα0.3

Legend: PINα0.3 | PINα0.5 | PINα0.7: Painting with light colour ($\alpha = 0.3$), medium ($\alpha = 0.5$) and dark ($\alpha = 0.7$), **GES1 | GES3**: Gypsum plaster with thicknesses of 1 and 3 cm, **ARG2 | ARG4 | ARG6**: Mortar with thicknesses of 2, 4 and 6 cm, **BCE9**: Masonry of 8-hole ceramic blocks horizontally 9 × 19 × 19 cm, **BCE14**: Masonry of ceramic blocks with vertical hole 14 × 19 × 39 cm, **BCO14**: Concrete block masonry with vertical hole 14 × 19 × 39 cm, **PCO10**: Solid concrete wall 10 cm thick

Note: Line 1 represents the reference model and cells filled in yellow represent the layer of the wall system changed from the reference model

2.3 Floor System

For the floor system that divides overlapping environments of the reference model was considered concrete slab with 7 cm thickness, plasterboard lining with thickness 2 cm and a distance to the concrete slab equal to 20 cm, without mineral wool or acoustic blanket, cemented mortar floor with 3 cm thickness and coating on light coloured ceramic boards.

Figure 7 presents schematic detail of the floor system considered in the reference model.

In addition to the floor system defined for the reference model, the following variations were adopted for the elements/components for the purpose of comparative analysis:

- No plaster lining;
- Inclusion of mineral wool with 5 cm thickness on the plaster lining;
- 3 different concrete slab thicknesses;
- Inclusion of 2 types of resilient blanket for acoustic flooring;
- 2 different floor thicknesses;
- 2 colours for floor covering.

Table 11 presents the relationship of floor systems, considering the variations mentioned above. Due to the geometric complexity and the non-homogeneous characteristics of some types of concrete slab for large spans, estimating the vibrational response in this type of floor requires methods to consider its different constructed forms (Oliveira and Patricio 2017). Given the complexity pointed out for the analysis of non-homogeneous concrete slabs, no floor system with ribbed or truss slabs was considered in the present study.

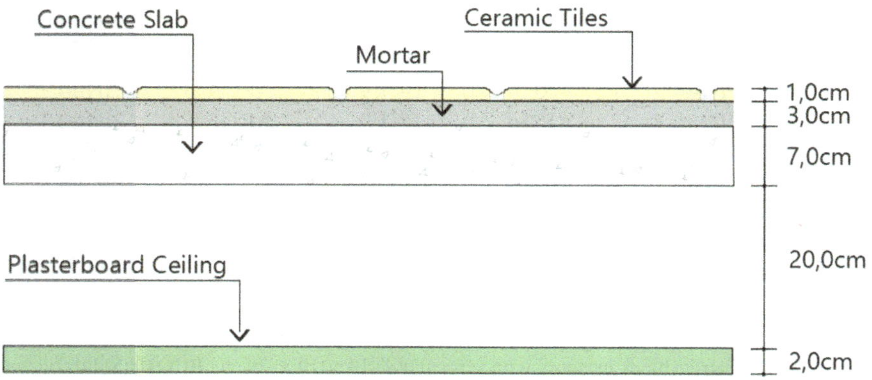

Fig. 7 Schematic detail of the reference floor system used in the study

Table 11 Floor systems (FS) considered in numerical simulations

FS	Lining		Structure	Sub-floor		Coating
1	FOR20	S/LÃ	LAJ7	S/MAC	ARG3	CERα0,3
2	S/FOR	S/LÃ	LAJ7	S/MAC	ARG3	CERα0.3
3	FOR20	LÃ5	LAJ7	S/MAC	ARG3	CERα0.3
4	FOR20	S/LÃ	LAJ10	S/MAC	ARG3	CERα0.3
5	FOR20	S/LÃ	LAJ13	S/MAC	ARG3	CERα0.3
6	FOR20	S/LÃ	LAJ16	S/MAC	ARG3	CERα0.3
7	FOR20	S/LÃ	LAJ7	MACΔ14	ARG3	CERα0.3
8	FOR20	S/LÃ	LAJ7	MACΔ29	ARG3	CERα0.3
9	FOR20	S/LÃ	LAJ7	S/MAC	ARG5	CERα0.3
10	FOR20	S/LÃ	LAJ7	S/MAC	ARG7	CERα0.3
11	FOR20	S/LÃ	LAJ7	S/MAC	ARG3	CERα0.5
12	FOR20	S/LÃ	LAJ7	S/MAC	ARG3	CERα0.7

Legend: FOR20: Plasterboard covering with distance of 20 cm for the concrete slab, **S/FOR:** Without lining, **S/LÃ:** No mineral wool, **LÃ5:** Mineral wool with thicknesses of 5 cm on the lining, **LAJ7 | LAJ10 | LAJ13 | LAJ16:** Concrete slab with thicknesses 7, 10, 13 and 16 cm, **SMAC:** No acoustic blanket, **MACΔ14:** Acoustic blanket with ΔLw equal to 14 dB, **MACΔ29:** Acoustic blanket with ΔLw equal to 29 dB, **ARG3 | ARG5 | ARG7**: Cement mortar with thicknesses of 3, 5 and 7 cm, **CERα0.3 | CERα0.5 | CERα0.7**: Ceramic boards with light colour ($\alpha = 0.3$), medium ($\alpha = 0.5$) and dark ($\alpha = 0.7$)

Note: Line 1 represents the reference model and cells filled in yellow represent the layer of the wall system changed from the reference model

2.4 Roof System

In the roof system of the reference model it was considered concrete slab with 7 cm thickness, plasterboard lining with thickness 2 cm, light colour paint and distance to the concrete slab equal to 20 cm, without mineral wool or thermal blanket, a waterproofing with asphalt blanket with thickness 4 mm, mechanical protection with cementitious mortar of 5 cm thickness and coating with light colour paint.

Figure 8 presents schematic detail of the floor system considered in the reference model.

In addition to the roof system defined for the reference model, the following variations were adopted for the elements/components for the objective of comparative analysis:

- 2 colours for the inner coating of the plaster lining;
- No plaster lining;
- Inclusion of mineral wool with 5 cm thickness on the plaster lining;
- 3 different concrete slab thicknesses;
- Inclusion of 2 types of thermal insulators on the structural element;
- Waterproofing replacement with asphalt blanket and mechanical roof protection with fiber cement tiles;
- 2 colours for the external coating of the roofing system.

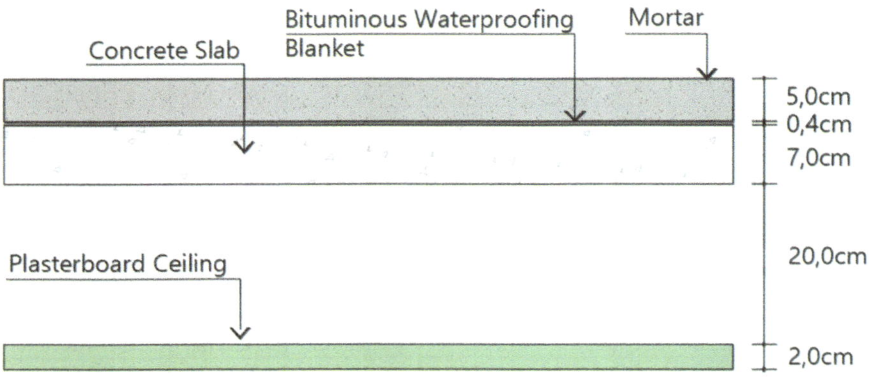

Fig. 8 Schematic detail of the reference roof system used in the study

Table 12 presents the relationship of roofing systems, considering the variations previously pointed out.

Table 12 Roofing systems for numerical simulations

FS	Lining		Structure	Sub-floor		Coating
1	FOR20α0.3	S/LÃ	LAJ7	S/MTE	IMP0.4 + ARG5	PINα0.3
2	FOR20α0.5	S/LÃ	LAJ7	S/MTE	IMP0.4 + ARG5	PINα0.3
3	FOR20α0.7	S/LÃ	LAJ7	S/MTE	IMP0.4 + ARG5	PINα0.3
4	S/FOR	S/LÃ	LAJ7	S/MTE	IMP0.4 + ARG5	PINα0.3
5	FOR20α0.3	LÃ5	LAJ7	S/MTE	IMP0.4 + ARG5	PINα0.3
6	FOR20α0.3	S/LÃ	LAJ10	S/MTE	IMP0.4 + ARG5	PINα0.3
7	FOR20α0.3	S/LÃ	LAJ13	S/MTE	IMP0.4 + ARG5	PINα0.3
8	FOR20α0.3	S/LÃ	LAJ16	S/MTE	IMP0.4 + ARG5	PINα0.3
9	FOR20α0.3	S/LÃ	LAJ7	XPS2	IMP0.4 + ARG5	PINα0.3
10	FOR20α0.3	S/LÃ	LAJ7	EPS4	IMP0.4 + ARG5	PINα0.3
11	FOR20α0.3	S/LÃ	LAJ7	S/MTE	TFC	PINα0.3
12	FOR20α0.3	S/LÃ	LAJ7	S/MTE	IMP0.4 + ARG5	PINα0.5
13	FOR20α0.3	S/LÃ	LAJ7	S/MTE	IMP0.4 + ARG5	PINα0.7

Legend: FOR20α0.3: Plasterboard lining with light colour paint ($\alpha = 0.3$) and distance of 20 cm to the concrete slab, **FOR20α0.5:** Plasterboard lining with medium colour paint ($\alpha = 0.5$) and distance of 20 cm to the concrete slab, **FOR20α0.7:** Plasterboard lining with dark colour paint ($\alpha = 0.7$) and distance of 20 cm to the concrete slab, **S/FOR:** No lining, **S/LÃ:** No mineral wood, **LÃ5:** Mineral wool thickness 5 cm on the lining, **LAJ7 | LAJ10 | LAJ13 | LAJ16:** Concrete slab with a thicknesses 7, 10, 13 and 16 cm, **S/MTE:** No thermal blanket, **XPS2:** Thermal blanket. XPS with 2 cm thickness, **IMP0.4 + ARG5:** Asphalt blanket 0.4 cm thick and mechanical mortar protection 5 cm, **EPS4:** Thermal blanket. EPS 4 cm thick, **TFC:** Roof with fiber cement tiles and tube larger than 5 cm, **PINα0.3 | PINα0.5 | PINα0.7**: Painting with light colour ($\alpha = 0.3$), medium ($\alpha = 0.5$) and dark ($\alpha = 0.7$)

Note: Line 1 represents the reference model and cells filled in yellow represent the layer of the wall system changed from the reference model

2.5 *Adopted Method for the Study of Acoustic Performance*

For acoustic performance analysis through numerical simulation, two specific software's were used: *Insul*, used to predict the sound reduction index (Rw) of the opaque elements, simulating the characterization by the precision method, laboratory assay, as recommended by the precision method of NBR 15575-3 (2013) and NBR 15575-4 (2013); and the *Sonarchitect ISO Professional*, used to verify the acoustic insulation of the walls in the field, following the parameters of international standards ISO 12354-1 (2017), ISO 12354-2 (2017), ISO 12354-3 (2017), simulating field trials by the engineering method, recommended in NBR 15575-3 (2013), NBR 15575-4 (2013) and NBR 15575-5 (2013).

2.5.1 Insul

Insul is a software created by Marshall Day Acoustic and aims to calculate the prediction of the sound insulation of a wall. This calculation is based on theoretical equations (mass theory, critical frequency and others) in order to assist designers in choosing the best materials for acoustic projects (INSUL 2019).

Insul is one of the main software used in building acoustics, more specifically for predicting sound insulation indexes (Rw) and standard impact sound pressure level (L'nT,w) for vertical wall systems, roofing systems or flooring systems in a simple and practical way, presenting errors of up to 3 dB in relation to laboratory measurements.

Insul offers several materials used in Europe with their predefined properties. However, there are few data in the acoustic area of materials conventionally used in Brazil, complicating the process of filling all the necessary characteristics for the operation of the software. Due to this limitation, it was necessary to create new materials and elements for the study.

In the present study, to create the elements in *Insul*, laboratory testes systems were used as reference with results made available by the Guide to meet the performance standard. The walls and their sound reduction index (Rw) results used as initial reference are presented in Tables 13 and 14.

The main input data requested by *Insul* for editing the materials are: thickness; density; Young modulus of elasticity and damping, as can be seen on the software material editing screen, shown in Fig. 9. The material properties considered as input data in *Insul* are presented in Table 15.

The sound reduction indexes obtained at *Insul* and used in numerical simulation are presented in Tables 16 and 17 for elements of vertical walls and floor systems, respectively.

With the acoustic characteristics of the materials admitted from the exposed methodology, each element provided for the study was modelled, presented in Tables 9, 10, 11 and 12. Figures 10 and 11 present examples of *Insul* software

Table 13 Indicative values of weighted sound reduction index for some wall systems

Type of Wall	Block/brick width	Coating	Approximate mass (kg/m^2)	Rw (dB)
Concrete hollow blocks	9 cm	Mortar with 1.5 cm on each face	180	41
	11.5 cm		210	42
	14 cm		230	45
Ceramic hollow blocks	9 cm	Mortar with 1.5 cm on each face	120	38
	11.5 cm		150	40
	14 cm		180	42
Solid bricks of baked clay	11 cm	Mortar with 2.0 cm on each face	260	45
	15 cm		320	47
	11 cm + 11 cm[a]		450	52
Walls of reinforced concrete	5 cm	No coating	120	38
	10 cm		240	45
	12 cm		290	47
Drywall	2 boards + glass wool	No coating	22	41
	4 boards		44	45
	4 boards + glass wool		46	49

[a]Double wall 11 + 11 cm, with 4 cm internal space filled with rock wool blanket of 70 kg/m^3

Table 14 Indicative values of weighted standardized impact sound pressure index, $L'_{nT,w}$

Type of product used in floating floor and concrete slab results without any acoustic treatment	Impact Sound Pressure Index (dB)
Concrete slab thickness with 10 cm, without resilient blanket and without sub-floor	82
Concrete slab thickness with 15 cm, without resilient blanket and without sub-floor	71
Blanket thickness 10 mm with synthetic rubber and 88% recycled material, without sub-floor	58
5 mm thick recycled rubber blanket (800 kg/m^3)—no flooring	58
Rubber blanket recycled thickness 3 mm (600 kg/m^3), plus 5 cm sub-floor	64
Synthetic wool blanket + sub-floor 5 cm	57
Polypropylene blanket with 10 mm + sub-floor 5 cm	52
Polypropylene blanket with 5 mm + sub-floor 5 cm	60

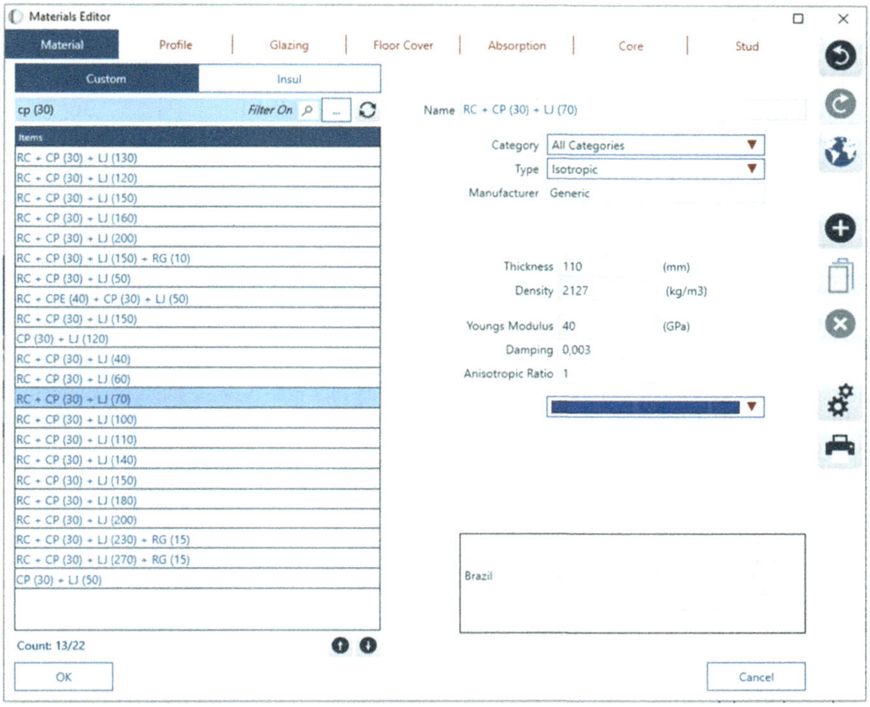

Fig. 9 *Insul* material editing screen

Table 15 Properties of materials used for modelling in *Insul* NBR 15220-2 (2005)/engineering noise control—Theory and pratice (Bies and Hansen 2009)

Material	Density (kg/m^3)	Modulus of elasticity (GPa)	Damping
Gypsum	1100	30	0.003
Mortar	1600	30	0.003
Concrete	2400	40	0.006
Ceramic block	616	10	0.011
Concrete block	896	40	0.001
Ceramic board	1600	4.68	4.68
Gypsum board	900	30	30
Glass wool	22	–	–

Note: The frames and acoustic blankets were not modelled in *Insul*, and the Rw and ΔLw were used directly in the *SONarchitect*

Table 16 Sound reduction indexes (Rw) of the elements of the vertical wall systems obtained at *INSUL*

IWS	Coating 1	Wall	Coating 1	Rw
1	GES1	BCE9	GES1	36
4	GES3	BCE9	GES3	39
5	ARG2	BCE9	ARG2	39
6	ARG4	BCE9	ARG4	43
7	ARG6	BCE9	ARG6	47
8	GES1	BCE14	GES1	40
9	GES1	BCO14	GES1	44
10	GES1	PCO10	GES1	46

Legend: GES1 | GES3: Gypsum plaster with thicknesses of 1 and 3 cm, **ARG2 | ARG4 | ARG6**: Mortar with thicknesses of 2, 4 and 6 cm, **BCE9**: Masonry of 8-hole ceramic blocks horizontally 9 × 19 × 19 cm, **BCE14**: Masonry of ceramic blocks with vertical hole 14 × 19 × 39 cm, **BCO14**: Concrete block masonry with vertical hole 14 × 19 × 39cm, **PCO10**: Solid concrete wall 10 cm thick

Note: Line 1 represents the reference model and cells filled in yellow represent the layer of the wall system changed from the reference model

Table 17 Sound reduction indexes (Rw) and sound pressure level of weighted standard impact of floor systems obtained at *INSUL*

FS	Lining		Structure	Sub-floor		Rw	L'nT,w
1	FOR20	S/LÃ	LAJ7	S/MAC	ARG3	59	70
2	S/FOR	S/LÃ	LAJ7	S/MAC	ARG3	50	81
3	FOR20	LÃ5	LAJ7	S/MAC	ARG3	61	67
4	FOR20	S/LÃ	LAJ10	S/MAC	ARG3	61	68
5	FOR20	S/LÃ	LAJ13	S/MAC	ARG3	62	67
6	FOR20	S/LÃ	LAJ16	S/MAC	ARG3	63	64
7	FOR20	S/LÃ	LAJ7	MACΔ14	ARG3	59	58
8	FOR20	S/LÃ	LAJ7	MACΔ29	ARG3	59	43
9	FOR20	S/LÃ	LAJ7	S/MAC	ARG5	60	69
10	FOR20	S/LÃ	LAJ7	S/MAC	ARG7	61	69

Legend: FOR20: Plasterboard covering with distance of 20 cm for the concrete slab, **S/FOR:** Without lining, **S/LÃ:** No mineral wool, **LÃ5:** Mineral wool with thicknesses of 5 cm on the lining, **LAJ7 | LAJ10 | LAJ13 | LAJ16:** Concrete slab with thicknesses 7, 10, 13 and 16 cm, **S/MAC:** No acoustic blanket, **MACΔ14:** Acoustic blanket with ΔLw equal to 14 dB, **MACΔ29:** Acoustic blanket with ΔLw equal to 29 dB, **ARG3 | ARG5 | ARG7**: Cement mortar with thicknesses of 3, 5 and 7 cm.

Note: Line 1 represents the reference model and cells filled in yellow represent the layer of the wall system changed from the reference model

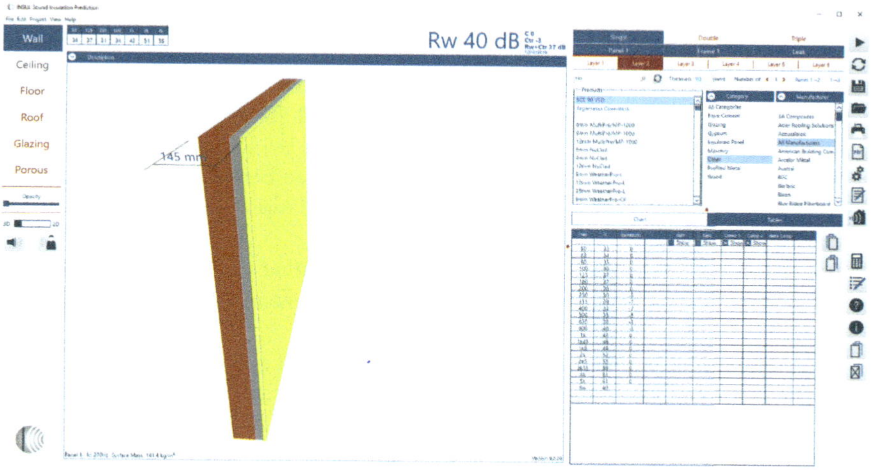

Fig. 10 *Insul* screen with wall system modelling

Fig. 11 *Insul* screen with floor system modelling

screens used in the study with modelling of external vertical wall systems and floor system, respectively.

2.5.2 Sonarchitect ISO Professional

The *SONarchitect ISO* was developed by Sound Of Numbers (SON) and is a tool for calculating sound insulation according to international standard ISO 12354 Parts 1 to 3 (2017).

The software allows the analysis of the design of an entire building, estimating the isolation to external air sound, internal sound insulation (impact and aerial) and reverberation time.

It is a software widely used by acoustics project offices and in academic works, being used, for example, in the researches of Silva, Rezende (2019) and TakahashI (2016).

In a comparative study, Remígio et al. (2019) found a good relationship between computational simulation performed with *SONarchitect* and infield trials, as of the 14 situations studied, 92.86% presented results within the 2 dB margin.

That is, the SONarchitect ISO is a tool for calculating sound transmission, air and impact noise, field wall systems, considering the connections between elements, marginal transmissions, building geometry and other parameters provided for in ISO 12354, representing a method of prediction of sound insulation measured in the field, by the engineering method, as recommended by NBR 15575-1 (2013).

The main input data used for computational simulation in SONarchitect ISO are the building geometry, in this case, considered the reference model presented in Fig. 7, and the values of sound reduction indexes of the systems obtained through computational modelling by the *Insul* software, as described in Sect. 2.5.1.

For the composition of the equivalent acoustic insulation index (Rweq) of the external vertical wall system (façade) was inserted in the *SONarchitect* the Rw of the opaque element, obtained by *Insul*, and the Rw of the window, defined in the reference model.

Although *SONarchitect* ISO allows the import of DXF files, which facilitates the analysis of building projects, this tool was not used for the present study, and the reference model is inserted in the *SONarchitect* ISO using the software's own design tools.

Figure 12 shows the floor plan of the reference model inserted in the *SONarchitect* ISO. It should be noted that the lines of the vertical and horizontal seals detailed in the software are relative to its axis, so for each variation of the thickness of the walls a model was created, in order to preserve the internal dimensions established in the reference model. For each variation of the wall systems, the following sound insulation parameters were recorded:

- **$D_{nT,w}$(IWS):** Standardized weighted level difference of internal wall;
- **$D_{2m,\ nT,w}$(EWS):** Standardized difference of weighted level 2 m away from the façade of the external wall (façade);
- **$D_{nT,w}$(FS):** Standardized weighted level difference of floor (Floor System);
- **$L'_{nT,w}$(FS):** Weighted standard impact sound pressure level of the floor system.

Figures 13 and 14 present *SONarquitect* ISO screens captured during the analysis of the standardized weighted level difference of the internal wall ($D_{nT,w}$ (IWS)) and

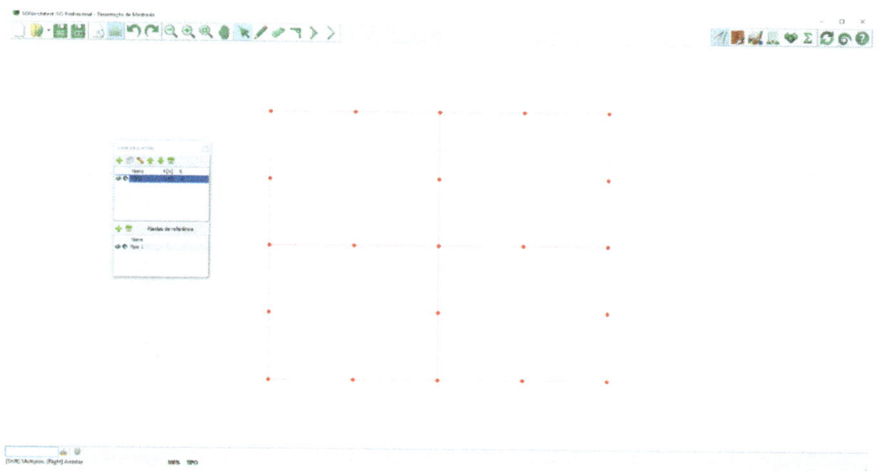

Fig. 12 Base plan of the reference model in *SONarchitect ISO*

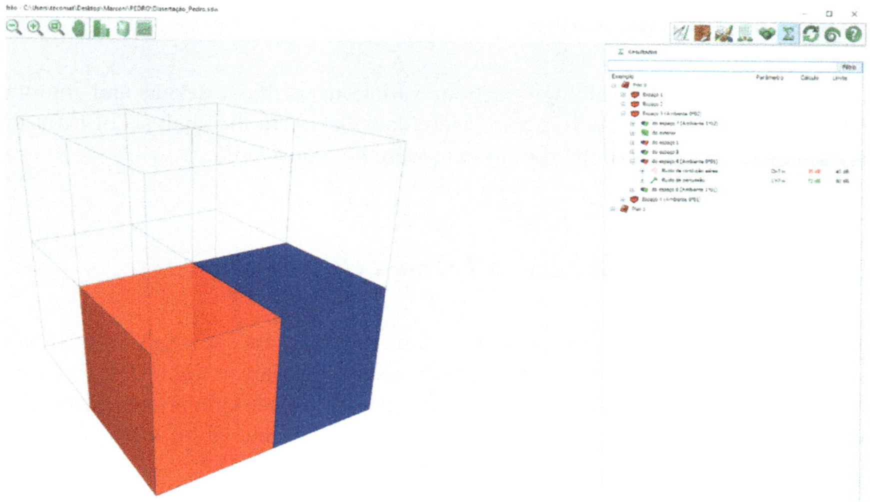

Fig. 13 *SONarchitect ISO* screen during analysis of the standardized weighted level difference of the internal vertical wall of the reference model

Standardized difference of weighted level at 2 m far away from the façade ($D_{2m,\ nT,w}$), respectively.

With all the defined situations modelled and the sound insulation values recorded for each model, the methodology of this research consisted of the comparative analysis of the results in relation to the reference model, whose results and discussions are presented in Sect. 3.

Fig. 14 *SONarchitect ISO* screen during sound insulation analyses

3 Results and Discussion

As defined in the Methodology section, variations in the external and internal wall systems and in the floor systems were considered. In the analysis of acoustic performance, variations in the roof system were not considered.

3.1 Acoustic Performance in Façades (EWS)

The variations related to external wall systems (façades) are shown in Table 18 and the results with the respective increase in acoustic insulation in Table 19.

The variations in colours were not considered in the simulations, given that this exchange does not interfere with anything in the simulation input data and, consequently, does not influence the final performance.

It is worth mentioning that in these models, the internal wall systems, floor systems and roof system of the reference model were considered, as Tables 9, 10 and 11, respectively.

From the analysis of the results presented in Table 19 it is possible to reach the following conclusions:

- The variation in the thickness of the internal and external coatings of the external wall (facade) does not increase the insulation of the external or internal walls (IDs 2 to 5, 9 and 10). They also do not show a significant increase in the insulation of the floor system or internal fences, with 1 dB being the largest recorded;

Table 18 External wall systems considered in numerical simulations

EWS	Internal coating		Wall	External coating		Window
1	PINα0.3	GES1	BCE9	ARG3	CERα0.3	ESQRW15FS0.85
4	PINα0.3	GES3	BCE9	ARG3	CERα0.3	ESQRW15FS0.85
5	PINα0.3	ARG2	BCE9	ARG3	CERα0.3	ESQRW15FS0.85
6	PINα0.3	ARG4	BCE9	ARG3	CERα0.3	ESQRW15FS0.85
7	PINα0.3	ARG6	BCE9	ARG3	CERα0.3	ESQRW15FS0.85
8	PINα0.3	GES1	BCE14	ARG3	CERα0.3	ESQRW15FS0.85
9	PINα0.3	GES1	BCO14	ARG3	CERα0.3	ESQRW15FS0.85
10	PINα0.3	GES1	PCO10	ARG3	CERα0.3	ESQRW15FS0.85
11	PINα0.3	GES1	BCE9	ARG5	CERα0.3	ESQRW15FS0.85
14	PINα0.3	GES1	BCE9	ARG7	CERα0.3	ESQRW15FS0.85
17	PINα0.3	GES1	BCE9	ARG3	TEXα0.3	ESQRW15FS0.85
18	PINα0.3	GES1	BCE9	ARG3	CERα0.3	ESQRW19FS0.66
19	PINα0.3	GES1	BCE9	ARG3	CERα0.3	ESQRW23FS0.52

Legend: PINα0.3: Painting with light colour ($\alpha = 0.3$), **GES1 | GES3**: Gypsum plaster with thicknesses of 1 and 3 cm, **ARG2 | ARG3 | ARG4 | ARG5 | ARG6 | ARG7**: Mortar with thicknesses from 2 to 7 cm, BCE9: Masonry of 8-hole ceramic blocks horizontally 9 × 19 × 19 cm, **BCE14**: Masonry of ceramic blocks with vertical hole 14 × 19 × 39 cm, **BCO14**: Concrete block masonry with vertical hole 14 × 19 × 39 cm, PCO10: Solid concrete wall 10 cm thick, **CERα0.3 | CERα0.5 | CERα0.7**: Ceramic boards with light, medium and dark colour, **TEXα0.3 | TEXα0.5 | TEXα0.7**: Acrylic texture with light, medium and dark colour, **ESQRW15FS0.85**: Window Rw = 15 dB and colourless Float glass 4 mm (Solar Factor = 0.85), **ESQRW19FS0.66**: Window Rw = 19 dB and green Float glass 4 mm (Solar Factor = 0.66), **ESQRW23FS0.52**: Window Rw = 23 dB and grey laminated glass 6 mm (Solar Factor = 0.52)

Note: Line 1 represents the reference model and cells filled in yellow represent the layer of the wall system changed from the reference model

- The variations in the structural element of the external wall did not increase the acoustic insulation of the internal or external wall systems. They also do not show a significant increase in the insulation of the floor system or internal fences, with 1 dB being the highest recorded (ID 6, 7 and 8);
- The variation of the external frame showed a significant increase, reaching an increase of 8 dB in relation to the reference model, in the acoustic insulation of the external walls (ID 12 and 13).

It can be concluded that the variations in the layers of finish, coating and even of the structural elements of the external wall, did not confer a significant increase in the acoustic insulation of the wall systems of the analysed model.

In contrast, the variation in the sound insulation index of the external frames proved to be the main interference variable in the acoustic insulation of the model's external wall, governing the acoustic performance of the facade.

Making a comparison with the minimum performance criteria recommended in NBR 15575-4 (2013), presented in Table 5, the reference model, with $D_{2m,nT,w} = 22$ dB, would only meet the noise class I, while considering the variations in the

Table 19 Sound insulation results obtained in the numerical simulation considering the variations in the external wall system (EWS)

ID	EWS	Variation	$D_{nT,w}$ (IWS) (dB)	$D_{2m,nT,w}$ (dB)	$D_{nT,w}$ (SP) (dB)	$L'_{nT,w}$ (dB)
			Result (Increm)*	Result (Increm)*	Result (Increm)*	Result (Increm)*
1	1	Reference model	35	22	49	76
2	4	Internal coating GES1 → GES3	35 (0)	22 (0)	49 (0)	76 (0)
3	5	Internal coating GES1 → ARG2	35 (0)	22 (0)	49 (0)	76 (0)
4	6	Internal coating GES1 → ARG4	35 (0)	22 (0)	49 (0)	75 (-1)
5	7	Internal coating GES1 → ARG6	35 (0)	22 (0)	49 (0)	75 (-1)
6	8	Wall BCE9 → BCE14	35 (0)	22 (0)	49 (0)	75 (-1)
7	9	Wall BCE9 → BCO14	35 (0)	22 (0)	50 (+1)	75 (-1)
8	10	Wall BCE9 → PCO10	35 (0)	22 (0)	50 (+1)	75 (-1)
9	11	External coating ARG3 → ARG5	35 (0)	22 (0)	49 (0)	75 (-1)
10	14	External coating ARG3 → ARG7	35 (0)	22 (0)	49 (0)	75 (-1)
11	17	External finishing CERα0,3 → TEXα0,3	35 (0)	22 (0)	49 (0)	76 (0)
12	18	Window ESQRW15→ESQRW19	35 (0)	26 (+4)	49 (0)	76 (0)
13	19	Window ESQRW15→ESQRW23	35 (0)	30 (+8)	49 (0)	76 (0)

ID: Identification of the numerical simulation
Note: Results that show positive increase in relation to the model highlighted with green filling
*Increment relative to the result obtained for the reference model

sound reduction indices (Rw) of the frames to 19 dB and 23 dB, the external wall system presented $D_{2m,nT,w} = 26$ dB and $D_{2m,nT,w} = 30$ dB, according to noise classes II and III, respectively.

In resume, considering the reference model, the main strategy for improving the acoustic insulation of the facade would be to improve the sound reduction index (Rw) of the external frames.

3.2 Acoustic Performance in Internal Wall Systems (IWS)

Variations related to internal wall systems (IWS) are shown in Table 20 and the results with the respective increase in acoustic insulation in the Table 21.

The variations in colours were not considered in the simulations, given that this exchange does not interfere with anything in the simulation input data and, consequently, does not influence the final performance.

It is worth mentioning that in these models the external wall systems, floor systems and roof system of the reference model were considered, as Tables 9, 10 and 11, respectively.

From the analysis of the results presented in the Table 21, we can reach the following conclusions:

- The increase in the thickness of the coatings provides an important increase in the acoustic insulation of the internal walls. It is noticed that such increase has a direct correlation with the increase in the thickness of the coatings (ID 14–17);
- The alteration of the structural element of the fence provided a relevant increase in the acoustic insulation to aerial noise of the internal wall system (ID 18 and 20).
- By comparing the BCE14 and BCO14, it can be seen that the increase in the density of the wall element has a direct correlation with the increase in sound insulation, thus, despite the same dimension, the concrete block has greater sound insulation;
- The changes in the IWS have no significant increase in the insulation of the floor system or external or floor seals, with 1 dB being the largest recorded.

It can be concluded that the increase in coatings and the use of structural elements with greater density were presented as the main devices for increasing the acoustic

Table 20 Internal walls systems considered in numerical simulations

IWS	Coating 1		Wall	Coating 2	
1	PINα0.3	GES1	BCE9	GES1	PINα0.3
4	PINα0.3	GES3	BCE9	GES3	PINα0.3
5	PINα0.3	ARG2	BCE9	ARG2	PINα0.3
6	PINα0.3	ARG4	BCE9	ARG4	PINα0.3
7	PINα0.3	ARG6	BCE9	ARG6	PINα0.3
8	PINα0.3	GES1	BCE14	GES1	PINα0.3
9	PINα0.3	GES1	BCO14	GES1	PINα0.3
10	PINα0.3	GES1	PCO10	GES1	PINα0.3

Legend: PINα0.3: Painting with light colour ($\alpha = 0.3$), **GES1 | GES3**: Gypsum plaster with thicknesses of 1 and 3 cm, **ARG2 | ARG4 | ARG6**: Mortar with thicknesses of 2, 4 and 6 cm, **BCE9**: Masonry of 8-hole ceramic blocks horizontally 9 × 19 × 19 cm, **BCE14**: Masonry of ceramic blocks with vertical hole 14 × 19 × 39 cm, **BCO14**: Concrete block masonry with vertical hole 14 × 19 × 39 cm, **PCO10**: Solid concrete wall 10 cm thick

Note: Line 1 represents the reference model and cells filled in yellow represent the layer of the wall system changed from the reference model

Table 21 Sound insulation results obtained in the numerical simulation considering the variations in the internal wall systems (IWS)

ID	IWS	Variation	$D_{n,Tw}$ (IWS) (dB)	$D_{2m,nT,w}$ (dB)	$D_{nT,w}$ (FS) (dB)	L'_{nTw} (dB)
			Result (Increment)*	Result (Increment)*	Result (Increment)*	Result (Increment)*
1	1	Reference model	35	22	49	76
14	4	Coating GES1+1→GES3+3	37 (+2)	22 (0)	49 (0)	76 (0)
15	5	Coating GES1+1→ARG2+2	37 (+2)	22 (0)	49 (0)	76 (0)
16	6	Coating GES1+1→ARG4+4	41 (+6)	22 (0)	49 (0)	76 (0)
17	7	Coating GES1+1→ARG6+6	44 (+9)	22 (0)	49 (0)	75 (-1)
18	8	Wall BCE9 → BCE14	38 (+3)	22 (0)	49 (0)	76 (0)
19	9	Wall BCE9→BCO14	42 (+7)	22 (0)	49 (0)	75 (-1)
20	10	Wall BCE9→PCO10	43 (+8)	22 (0)	49 (0)	75 (-1)

ID: Identification of the numerical simulation
Note: Results that show positive increase in relation to the model highlighted with green filling
*Increment relative to the result obtained for the reference model

insulation of the internal wall system of the reference model. This verification can be explained by the law of the masses.

Making a comparative analysis in relation to the criteria of the Performance Standard, NBR 15575-1 (2013), presented in Table 7, it appears that the reference model would not meet the minimum performance level, even if none of the environments is a dormitory ($D_{n,Tw} \geq 40$ dB).

Only considering the alteration of the coating for mortar with a thickness of 4 cm on each side or the change of the structuring element for masonry of concrete blocks 14 × 19 × 39 cm, or solid concrete wall with a thickness of 10 cm, was it found results of insulation to aerial noise greater than 40 dB.

Still considering the minimum performance criteria of NBR 15575-1 (2013), presented in Table 7, none of the results obtained for the reference model would be sufficient to meet the requirement if one of the environments was a bedroom.

In short, we can see that, for the variations considered in the reference model, the main vector for increasing the acoustic insulation of the internal wall system is the increase in the mass of its elements.

In addition, even considering the reference model, it is possible to verify that the vertical fencing systems conventionally used for construction of residential buildings have low potential to meet the minimum performance level prescribed by the performance standard, especially when one of the environments is a bedroom.

3.3 Acoustic Performance in Floor System (FS)

Variations related to floor systems (FS) are shown in Table 22 and the results with the respective increase in acoustic insulation in Table 21.

The variation of colours were not considered in the simulations, considering that such exchange does not interfere with anything in the simulation input data and, consequently, does not influence the final performance.

It should be noted that in these models, the external and internal wall systems and the reference model roofing system were considered, as in Tables 18, 19 and 21, respectively.

From the results presented in Table 23, we can do the following analyses:

- The removal of the ceiling considered in the reference model caused a significant decrease in the acoustic insulation both to aerial noise and to the impact noise of the floor system and still had a less significant impact, reducing the acoustic insulation of the internal wall system (ID 21);

Table 22 Floor systems considered in numerical simulations

FS	Lining		Structure	Sub-floor		Coating
1	FOR20	S/LÃ	LAJ7	S/MAC	ARG3	CERα0.3
2	S/FOR	S/LÃ	LAJ7	S/MAC	ARG3	CERα0.3
3	FOR20	LÃ5	LAJ7	S/MAC	ARG3	CERα0.3
4	FOR20	S/LÃ	LAJ10	S/MAC	ARG3	CERα0.3
5	FOR20	S/LÃ	LAJ13	S/MAC	ARG3	CERα0.3
6	FOR20	S/LÃ	LAJ16	S/MAC	ARG3	CERα0.3
7	FOR20	S/LÃ	LAJ7	MACΔ14	ARG3	CERα0.3
8	FOR20	S/LÃ	LAJ7	MACΔ29	ARG3	CERα0.3
9	FOR20	S/LÃ	LAJ7	S/MAC	ARG5	CERα0.3
10	FOR20	S/LÃ	LAJ7	S/MAC	ARG7	CERα0.3

Legend: FOR20: Plasterboard covering with distance of 20 cm for the concrete slab, **S/FOR:** Without lining, **S/LÃ:** No mineral wool, **LÃ5:** Mineral wool with thicknesses of 5 cm on the lining, **LAJ7 | LAJ10 | LAJ13 | LAJ16:** Concrete slab with thicknesses 7, 10, 13 and 16 cm, **S/MAC:** No acoustic blanket, **MACΔ14:** Acoustic blanket with ΔLw equal to 14 dB, **MACΔ29:** Acoustic blanket with ΔLw equal to 29 dB, **ARG3 | ARG5 | ARG7**: Cement mortar with thicknesses of 3, 5 and 7 cm, **CERα0.3 | CERα0.5 | CERα0.7**: Ceramic boards with light colour ($\alpha = 0.3$), medium ($\alpha = 0.5$) and dark ($\alpha = 0.7$)

Note: Line 1 represents the reference model and cells filled in yellow represent the layer of the wall system changed from the reference model

Table 23 Sound insulation results obtained in the numerical simulation considering the variations in the floor system (FS)

ID	FS	Variation	$D_{nT,w}$ (dB) (IWS)	$D_{2m,nT,w}$ (dB)	$D_{nT,w}$ (dB) (FS)	L'_{nTw} (dB)
			Result (Increment)*	Result (Increment)*	Result (Increment)*	Result (Increment)*
1	1	Reference model	35	22	49	76
21	2	Plaster lining FOR20→ S/FOR	34 (-1)	22	44 (-5)	85 (+9)
22	3	Plaster lining S/Lã→LÃ5	35	22	49	75 (-1)
23	4	Structural element LAJ7→LAJ10	35	23 (+1)	50 (+1)	74 (-2)
24	5	Structural element LAJ7→LAJ13	35	23 (+1)	52 (+3)	72 (-4)
25	6	Structural element LAJ7→LAJ16	35	23 (+1)	54 (+5)	71 (-5)
26	7	Acoustic blanket S/MAC→MACΔ14	35	22	49	64 (-12)
27	8	Acoustic blanket S/MAC→MACΔ29	35	22	49	49 (-27)
28	9	Sub-floor ARG3→ARG5	35	23 (+1)	49	75 (-1)
29	10	Sub-Floor ARG3→ARG7	35	23 (+1)	50 (+1)	74 (-2)

ID: Identification of the numerical simulation
Note: Results that show negative increment in relation to the model highlighted with orange filling and the results that show positive increase in relation to the model highlighted with green filling
*Increment relative to the result obtained for the reference model

- The use of mineral wool on the plaster lining provided an increase of only 1 dB in the insulation to noise from impacts of the floor system, and no increase in the acoustic insulation to aerial noise, showing itself as an ineffective solution (ID 22);
- The increase in the thickness of the structural element of the reference model resulted in a positive increase in acoustic insulation for both aerial and impact noise, with a direct correlation between the increase in thickness and the increase in acoustic insulation (IDs 23, 24 and 25);
- The increase in the thickness of the structural element also generated a minor increase in the acoustic insulation of the external wall system (IDs 23, 24 and 25);
- The adoption of the acoustic blanket between the structural element and the subfloor in the reference model showed a very high increase in terms of acoustic insulation to noise from impacts of the floor system (IDs 26 and 27). However, it did not confer any change in the insulation to aerial noise;

- The results presented in the simulation for the floor system without lining (ID 21) and for the floor covering system with ΔLw equal to 29 dB (ID 27) are compatible with the results obtained in tests presented in a scientific article by Zuchetto et al. (2016);
- The increase in the thickness of the subfloor gave an insignificant increase in relation to the acoustic insulation to impact noise and aerial noise of the floor system (IDs 28 and 29);
- The increase in the thickness of the subfloor also gave a minor increase in relation to the acoustic insulation of the external wall system of the reference model (IDs 28 and 29).

It can be seen that the changes in the floor system of the reference model did not have a significant impact on the internal or external wall systems.

From the results presented, the use of ceiling and the increase of the thickness of the structural element can be considered as the main strategies for increasing the insulation to aerial noise of the floor system of the reference model.

Still regarding the isolation to aerial noise of the floor system, in all the evaluated variations, except the floor model of the reference model without the ceiling (ID 21), they meet the minimum level of performance established in NBR 15575-3 (2013) and presented in Table 8, in the situation where one of the environments is a dormitory.

Compliance with the minimum level of performance related to the isolation of impact noise established in the Performance Standard is also met for most situations, except for the floor system of the reference model without the ceiling (ID 21), considering the required criterion floor system between autonomous units, as shown in Table 9.

However, if we consider the floor system of common area of prolonged use over housing unit, exposed in the same Table 9, only the system with the use of an acoustic blanket with Weighted Reduction of the Impact Sound Pressure Level (ΔLw) equal to 29 dB would meet the minimum level recommended in NBR 15575-3 (2013).

That is, in the case of long-term common areas, the increase in the thickness of the structural element or the subfloor, or even the use of acoustic blankets with a low reduction in the impact sound pressure will probably not be enough to guarantee the minimum acoustic performance established in the NBR 15575-3 (2013).

In short, considering the study carried out in the reference model, the best interventions for increasing the isolation from aerial noise are the use of ceiling tiles and the increase in the thickness of the structural element.

To increase the insulation to impact noise, the best strategy is to use an acoustic blanket followed by increasing the thickness of the structural element.

3.4 Acoustic Performance—Level of Influence of Simulated Variations

From the results obtained, presented in the Tables 18, 20 and 22, and considering all simulated variations, a qualitative scale was defined to classify the level of influence of the coating layers on the acoustic performance, presented in Table 24.

Taking as a reference the qualitative rating scale in the level of influence presented in Table 24 and the results obtained in the acoustic performance simulations presented in Tables 19, 21 and 23, it is possible to make the classification as shown in Table 25.

Table 24 Qualitative scale for classifying the level of influence of the coating layers on acoustic performance

Increment (i)*	Level of influence
$i \leq \pm 2dB$	Low
$\pm 2dB < i \leq \pm 5dB$	Medium
$i > \pm 5dB$	High
* *Increase relative to the result of the reference model*	

*Increase relative to the result of the reference model

Table 25 Level of influence of the variations adopted in the acoustic performance in relation to the reference model

System	System Variation	$D_{2m,nT,w}$ (EWS)	$D_{nT,w}$ (IWS)	$L'_{nT,w}$ (Floor)	$D_{nT,w}$ (Floor)
EWS	Internal coating thickness	Low	Low	Low	Low
	Wall structuring element	Low	Low	Low	Low
	External coating thickness	Low	Low	Low	Low
	Thickness of external coating layer	Low	Low	Low	Low
	Different external coatings	Low	Low	Low	Low
	Windows	High	Low	Low	Low
IWS	Coating thickness	Low	High	Low	Low
	Wall structuring element	Low	High	Low	Low
FS	Use of lining	Low	Low	High	High
	Mineral wool on the lining	Low	Low	Low	Low
	Concrete slab thickness	Low	Low	Medium	Medium
	Acoustic blanket	Low	Low	Low	High
	Sub-floor thickness	Low	Low	Low	Low

4 Conclusions

The Brazilian standard NBR 15575 "Housing Buildings—Performance" is under review in order to be published at the end of 2020, as a new version. In accordance with this, the present work intends to help the Brazilian decision-makers and give an applied and helpful guide for designers.

Considering the responsibility of the designers to correctly specify materials to be used in the new constructions in order to meet the performance levels established in the Performance Standard—NBR 15575-1 (2013)—and by the developer, according to the Performance Profile of the Enterprise, the study presented a summary of the influence on acoustic performance for each variation tested in the reference model, as shown in Table 25.

In summary, considering the reference model, the methodology adopted and the main analyses, the following conclusions were obtained:

1. Acoustic performance of façades
    - The coating systems have low influence on the acoustic performance of the façade, and external frames are the main element of influence in this requirement.
2. Acoustic performance of internal walls
    - Coating systems have a high influence on the acoustic performance of internal walls, being of the same order of magnitude as the influence of the façades. However, the largest increments were obtained with high thicknesses, from 4 cm on each side, which may not be feasible from an executive point of view.
3. Acoustic performance of floor systems
    - The use of liners in the floor system has provided an important increase in insulation for both air and impact noise. On the other hand, the use of mineral wool has not been shown to be an efficient solution for sound insulation.
    - The increase in the thickness of the structural layer of the floor system provided an increase in acoustic insulation for both air and impact noise. However, increasing the thickness of the floor proved to be an ineffective solution.
    - The use of resilient acoustic blankets in a floating floor system has proved to be the most efficient solution for increasing insulation to impact noise, although the solution does not present a relevant increase in air noise insulation.

In resume, the most important ideas to the scientific community, decision-makers, engineers, and academics can be expressed in "guidelines" to improve the acoustic performance:

- To improve the acoustic insulation of external vertical seals (facades), considering the construction systems studied here, the best strategy is to improve the insulation of the frames, being the optimization obtained mainly by using better components and glasses with larger thicknesses. To increase internal vertical seals or walls between buildings acoustic insulation, it is necessary to use heavier building systems (masses law) or that use the mass-spring-mass law, such as drywall systems or double walls. To increase noise isolation from the floor system impacts, percussion, the most efficient way identified was the use of resilient blankets under the floor. There are several materials available on the market for this purpose, from bituminous products, such as rubbers, polymers, such as expanded polypropylene, to natural material blankets or cork, for example. The most important thing is to evaluate the impact sound loss transmission appropriate to the project.

References

Bies DA, Hansen CH (2009) Engineering noise control - Theory and pratice, 4th edition. Spon Press, UK

Borges CAM (2008) The concept of building performance and its importance for the civil construction sector in Brazil. University of São Paulo, Brazil, São Paulo

Chan ALS (2012) Effect of adjacent shading on the thermal performance of residential buildings in a subtropical region. Appl Energy 92:516–522

Fringuellino M, Smith RS (1999) Sound transmission through hollow brick walls. Building Acoust 6(3):211–224

Gomes JEV (2015) Building performance evaluation according to ABNT NBR 15575: adaptation to the case of rehabilitated building. MSc Thesis, Faculdade de Engenharia da Universidade do Porto, Portugal

INSUL (2019) P.S.I. About Insul. Available in. Accessed in 24 nov 2019

ISO 6241 (1984) Performance standards in building—principles for their preparation and factors to be considered. International Organization for Standardization, Switzerland

ISO 12354-1 (2017) Building acoustics—estimation of acoustic performance of buildings from the performance of elements—Part 1: airborne sound insulation between rooms

ISO 12354-2 (2017) Building acoustics—estimation of acoustic performance of buildings from the performance of elements—Part 2: impact sound insulation between rooms

ISO 12354-3 (2017) Building acoustics—estimation of acoustic performance of buildings from the performance of elements—Part 3: airborne sound insulation against outdoor sound

Lorensi LS (2015). Critical analysis and propose advances in methodologies of the experimental performance tests to analysis based on ABNT NBR 15575 (2013). UFRS, 2015 NBR 15220–3: thermal performance in buildings Part 3: Brazilian bioclimatic zones and building guidelines for low-cost houses. Rio de Janeiro-RJ, Brazil

Manso MA, Mitidieri Filho CV (2007) Model of project management and coordination system for construction companies and developers in the city of São Paulo. Project Management & Technology

Melo MJAC (2007) Analysis of reports issued on "coffin-type buildings" in the metropolitan reef region: causes pointed to the landslides and interdictions. MSc Thesis, Brazil

NBR 5674 (2012) Building maintenance—requirements for maintenance management system. Rio de Janeiro-RJ, Brazil

NBR 6136 (2016) Hollow concrete blocks for concrete masonry—requirements. Rio de Janeiro-RJ, Brazil

NBR 15575-1 (2013) Residential buildings—performance Part 1: general requirements. Rio de Janeiro-RJ, Brazil

NBR 15575-1 (2017) Residential buildings—performance Part 1: general requirements. Rio de Janeiro-RJ, Brazil
NBR 15575-2 (2013) Residential buildings—performance Part 2: requirements for structural systems. Rio de Janeiro-RJ, Brazil
NBR 15575-3 (2013) Residential buildings—performance Part 3: requirements for floor systems. Rio de Janeiro-RJ, Brazil
NBR 15575-4 (2013) Residential buildings—performance Part 4: requirements for internal and external wall systems. Rio de Janeiro-RJ, Brazil
NBR 15575-5 (2013) Residential buildings—performance Part 5: requirements for roofing systems. Rio de Janeiro-RJ, Brazil
NBR 15575-6 (2013) Residential buildings—performance Part 6: requirements for hydro sanitary systems. Rio de Janeiro-RJ, Brazil
Oliveira MF, Patricio JV (2017) Impact noise of non-homogeneous floors: analysis of different input parameters for computational modeling predictions. J Civ Eng Archit 11(3):274–281
Proacústica (2017) Proacoustic manual for noise class of housing buildings, Rio de Janeiro, Brazil
Remígio EG, Barbosa MM, Verissimo JV, Junior OJ, Gois PF, Costa e Silva, AJ (2019) Comparative study between field tests and computer simulations for the isolation of aerial noise from VVI consisting of an 8-hole ceramic block. In: Proceedings of the XV ENCAC, João Pessoa, Brazil, p. 227–234 (in Portuguese)
Rezende ECL (2019) Comparative analysis of numerical simulations and field measurements of acoustic performance criteria in vertical internal seals and horizontal seals. MSc Thesis, UNICAP, Brazil
Souza JLP, Kern AP, Tutikian BF (2018) Quantitative and qualitative analysis of the performance standard (NBR 15575/2013) and main challenges of implementing higher education in residential multi-floor building. Project Manage Technol 13(1):127
Takahashi VFM (2016) Acoustic performance of buildings: computational tool of evaluation. PhD Thesis, UNICAMP, Brazil

Influence of the Coating System on the Thermal Performance of Buildings

J. M. P. Q. Delgado, P. Freitas Gois, A. J. Costa e Silva, A. C. Azevedo, Matheus Barbosa, and D. Queiroz

Abstract This work presents an extensive numerical simulation to analyse the influence of the coating layers on the performance of construction systems, in order to make the constructions projects feasible, not only economically, but also technically. Through computer simulations based on a defined reference model for the study, the present work study the influence of different layers of floor, roof and internal and external wall systems, on the thermal performance. Finally, analysing the results, for the reference model used, identified the materials and elements with the greatest influence on a thermal - all elements of tension, roofing system and facades, especially an absence of external cladding and the use of thermal blankets on the roof.

Keywords Building performance · Thermal performance · Coatings · Numerical simulations

---

J. M. P. Q. Delgado (✉) · A. C. Azevedo
Civil Engineering Department, CONSTRUCT-LFC, Universidade do Porto, Rua Dr. Roberto Frias, s/n, 4200-465 Porto, Portugal
e-mail: jdelgado@fe.up.pt

A. C. Azevedo
e-mail: antonio.costaazevedo@fe.up.pt

P. F. Gois · A. J. C. Silva
Civil Engineering Department, Universidade Católica de Pernambuco, Recife, Brazil
e-mail: pedrodefg@gmail.com

A. J. C. Silva
e-mail: angelo.silva@unicap.br

M. Barbosa · D. Queiroz
TECOMAT Engenharia, Rua Serra da Canastra, 391, Cordeiro, Recife, Pernambuco 50640-310, Brazil
e-mail: matheusmendonca.b@gmail.com

D. Queiroz
e-mail: deborahnqueiroz@gmail.com

J. M. P. Q. Delgado (ed.), *Efficient and Suitable Construction*, Building Pathology and Rehabilitation 17, https://doi.org/10.1007/978-3-030-62829-1_3

1 Introduction

The relation between man and construction, especially housing construction, is remote and intertwined with the evolutionary history of humanity and society itself. The establishment of functional requirements for buildings and their parts stems from the obvious premise that buildings, being indispensable to the life and activity of man, must have characteristics that correspond to and meet human needs. In other words, the establishment of functional requirements for buildings is a performance prescription (Gomes 2015).

The word performance is widely used throughout society and has quite broad meaning. It is used for hardware evaluation, professional analysis business and sports, for example. It is common to use performance to compare professionals and equipment's; in general, a desirable standard is defined, often informally, for comparison with the performance delivered. A more modern view of performance began to be structured in the twentieth century, where studies were proposed by the National Bureau of Sciences (NBS) during the 1920s. In the 1930s and 1940s, the first performance standards were developed and the English expression performance requirements emerged (Lorensi 2013).

After the Second World War and the consequent need to build large-scale buildings in the reconstruction movement, especially in Europe, the application of innovative construction technologies and systems at the time caused the incidence of high cases of pathological manifestations, generating high economic and social burdens. Given this scenario, the need for a more careful analysis of the performance of the construction systems used proved to be very relevant. From the end of the 1960s, the USA and some European countries devoted themselves to deepening their studies and striving to solidify the application of the concept of performance to buildings. In the last years there has been an increase in public awareness about the effects of the indoor environment on people's comfort and health. Besides the thermal environment, the indoor environment also includes indoor air quality, as well as acoustic and luminous environments (Bluyssen 2008; Mendell 2003).

For example, related to temperature and relative humidity, the thermal environment affects occupants' sensation and is considered to be the environmental factor most valorised by the occupants. Extensive studies have been conducted on thermal comfort, resulting in many thermal comfort equations. The PMV (Predicted Mean Vote) and PPD (Predicted Percentage Dissatisfied) based on Fanger's comfort equation are widely used in design guides and standards (ANSI/ASHRAE 55 2004; Bsen Iso 7730 1995). To establish an acceptable indoor environment, all these factors should be considered (Clausen and Wyon 2008). ASHRAE Standards 55 and 62 address different environmental factors. More recently, the equivalence of the discomfort caused by different physical qualities was examined. An equivalence of acoustic sensation to thermal sensation for short-term exposure was established. Specifically, a change in temperature of 1 °C had the same effect as a change in noise of 2.6 dB (Goldman 1999). Physical environmental parameters are all interrelated, and the feeling of comfort is a composite state involving an occupant's sensations of all these factors (Goldman 1999; Haghighat and Donnini 1999; Nagano and Horikoshi 2005; Eduardo et al. 2004).

ISO 6241 (1984) "Performance standards in building - Principles for their preparation and factors to be considered", published in 1984, is an important regulatory framework for building performance. ISO 6241 establishes general principles for the development of performance standards in civil construction expressed in functional requirements of users, linking the performance of buildings and user requirements. The objective of this Standard was to assist ISO signatory countries in the elaboration of Performance Standards, and to serve as a guide for the selection of requirements that can be applied in each case when talking about building performance (Borges 2008). Table 1 presents the requirements of users considered in ISO 6241 (1984).

Published in 1984, ISO 6241 (1984) is still a valid and important reference for defining the performance requirements of buildings, and perhaps its main gap in relation to contemporary requirements is sustainability, considering that, at the time, the theme did not have the current relevance. As in the European post-war period (1970s and 1980s), in Brazil, the construction of large-scale buildings induced the use of new techniques and constructive technologies. During this period the productivity was prioritized, without clear technical criteria for evaluating the innovations adopted to enable this productivity. As an example of this process of technological innovations without a more detailed performance analysis, is the cases of the "coffin buildings" built in the Metropolitan Region of Recife.

In 2000, a Study Committee and Working Groups were created with the objective of coordinating the discussion on the performance of buildings in the technical environment, seeking consensus for the development of a Brazilian standard NBR 15575 (NBR 15575-1 2013; NBR 15575-3 2013; NBR 15575-4 2013; NBR 15575-5 2013), within the scope of ABNT. In Brazil, the popular standard NBR 15575 (NBR 15575-1 2013; NBR 15575-3 2013; NBR 15575-4 2013; NBR 15575-5 2013) "Housing Buildings - Performance", is under review in order to be published, in 2020, a new version. According to NBR 15575-1 (NBR 15575-1 2013), the building performance is the "behaviour in use of a building and its systems". This definition makes clear the concept of scope for all housing buildings, regardless of the construction systems, elements and components used, because the object of the standard is the user behaviour in the building and its parts. This bias is different from most ABNT technical standards related to civil construction, which focuses on the prescription of methods of sizing and execution of specific components, elements and construction systems. The requirement does not express values, being naturally qualitative.

The main objective of NBR 15575 (NBR 15575-1 2013; NBR 15575-3 2013; NBR 15575-4 2013; NBR 15575-5 2013) is to establish the requirements and performance criteria applicable to housing buildings as an integrated completely, as well as to be evaluated in insulation for one or more specific systems, represented by each part of the standard. The definition of the performance requirements of NBR 15575 (NBR 15575-1 2013; NBR 15575-3 2013; NBR 15575-4 2013; NBR 15575-5 2013) represents the user's requirements.

Table 1 Users requirements

Category	Examples
1. Stability requirements	Mechanical resistance to static and dynamic actions, both individually and in combination. Resistance to impacts, intentional or unintentional abusive actions, accidental actions, cyclical effects
2. Fire safety requirements	Risks of fire and fire diffusion, respectively. Psychological effects of smoke and heat Alarm triggers time (detection and alarm systems). Building evacuation time (exit routes). Survival time (compartmentalization of fire)
3. Safety risks in use	Safety related to aggressive agents (protection against explosions, burns, sharp points and edges, moving mechanisms, electrical discharges, radioactivity, contact or inhalation of poisonous substances, infection. Safety during movement and movement (limitation of slipping on floors, unobstructed roads, handrails, etc.). Security against the improper entry of persons and/or animals
4. Sealing requirements	Water sealing (rain, subsoil, drinking water, wastewater, etc.). Air and gas seal. Dust and snow seal
5. Thermal and moisture requirements	Control of air temperature, thermal radiation, air speed and relative humidity (limitation of variation in time and space, response of controls). Condensation controls
6. Air purity requirements	Ventilation. Odour control
7. Acoustic requirements	Control of internal and external noise (continuous and/or intermittent). Sound intelligibility. Reverberation time
8. Visual requirements	Natural and artificial lighting (necessary lighting, stability, luminous contrast and very strong light protection. Sunlight; Possibility of darkness. Aspects of spaces and surfaces (colour, texture, regularity, levelling, verticality, horizontality, etc.). Visual contact, internally and with the outside world (threads and barriers related to privacy, protection against optical distortion)
9. Tactile requirements	Properties of surfaces, roughness, dryness, heat, elasticity. Protection against static electricity discharges
10. Dynamic requirements	Limitation of vibrations and accelerations of the whole set (transient and continuous). Convenience of pedestrians in areas exposed to the wind. Ease of movement (slope of ramps, arrangement of stairs steps). Room for manoeuvre (handling of doors, windows, control over equipment, etc.)

(continued)

Table 1 (continued)

Category	Examples
11. Hygiene requirements	Installation for care and hygiene of the human body. Water supply. Cleaning conditions. Release of wastewater, materials served and smoke. Limitation of contaminant emission
12. Requirements for specific uses spaces	Quantity, size, geometry, subdivision, and interrelation of spaces. Services and equipment. Conditions (capacity) of furniture and flexibility
13. Durability requirements	Conservation (permanence) of performance in relation to the necessary service life subject to regular maintenance
14. Economic requirements	Maintenance, operational and capital costs. Demolition costs

It is important to highlight that Brazil has continental dimensions, with a total area that corresponds to about 80% of the entire territory of Europe. As a result, in addition to climate issues, there is a great diversity of customs and cultures, which forces designers to decide on different finishing solutions for the façades, namely types and sizes of materials, predominant colours, etc.; depending on the region where the building will be constructed. However, when it is governed by the same regulation [NBR 15575 (NBR 15575-1 2013; NBR 15575-3 2013; NBR 15575-4 2013; NBR 15575-5 2013)] to be used throughout the country, it is very important to know the difference in performance between the various solutions available on the market so that it is possible to ensure, both for the designers and for the users, the expected technical and aesthetic result.

1.1 Performance Standard—NBR 15575

The popularly known Performance Standard is a technical standard of ABNT—Brazilian Association of Technical Standards, NBR 15575—Housing Buildings—Performance. It is noteworthy that NBR 15575 is divided into 6 parts, the first of which deals with the general requirements of the building and the following deals with the requirements of its parts, represented by systems. This division makes clear the concept of analysis of the overall analysis of the building and the parts that compose it. So the six parts of NBR 15575 are:

- NBR 15575-1: General requirements (Brazilian Association of Technical Standards. NBR 15575-1: Housing buildings—Performance. Part 1: General Requirements. Rio de Janeiro, 2013).
- NBR 15575-2: Requirements for structural systems (Brazilian Association of Technical Standards. NBR 15575-2: Housing buildings—Performance. Part 2: Requirements for structural systems. Rio de Janeiro, 2013).

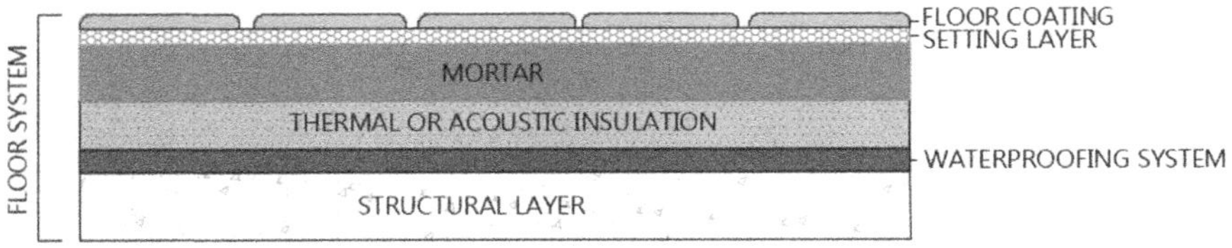

Fig. 1 Generic example of a floor system and its elements. Source NBR 15575-4 (2013)

- NBR 15575-3: Requirements for flooring systems (Brazilian Association of Technical Standards.. NBR 15575-3: Housing buildings—Performance. Part 3: Requirements for flooring systems. Rio de Janeiro, 2013).
- NBR 15575-4: Requirements for internal and external wall systems—SVVIE (Brazilian Association of Technical Standards. NBR 15575-4: Housing buildings—Performance. Part 4: Requirements for internal and external wall systems. Rio de Janeiro, 2013).
- NBR 15575-5: Requirements for roofing systems—SC (Brazilian Association of Technical Standards. NBR 15575-5: Housing buildings—Performance. Part 5: Requirements for roofing systems. Rio de Janeiro, 2013).
- NBR 15575-6: Requirements for hydro sanitary systems (Brazilian Association of Technical Standards. NBR 15575–6: Housing buildings—Performance. Part 6: Requirements for hydro sanitary systems. Rio de Janeiro, 2013).

It can be observed by dividing the parts of NBR 15575 that the performance analysis is specified for the building and its systems. Therefore, understanding the concept of system is very important and the standard itself defines system as "most functional part of the building. Set of elements and components intended to meet a macro-function that defines it."

Figure 1 exemplifies a floor system, where it can be intuitively identified that the final performance of the system will depend on the characteristic of the components and elements that compose it.

According to the definition of NBR 15575-1 (2013), performance is the "behaviour in use of a building and its systems". This definition makes clear the concept of scope for all housing buildings, regardless of the construction systems, elements and components used, because the object of the Standard is behaviour in use of the building and its parts. This bias is different from most ABNT technical standards related to civil construction, which focuses on the prescription of methods of sizing and execution of specific components, elements and construction systems.

The performance definition of NBR 15575 also explains the intimate relationship with the requirements of users and the exposure conditions to which the building is exposed, and the main objective of NBR 15575 is to establish requirements and performance criteria applicable to housing buildings, as an integrated whole, as well as to be evaluated in insulation for one or more specific systems, represented by each part of the Standard.

Performance requirements, according to the Standard itself, represent the conditions that qualitatively express the attributes that the housing building and its systems must have, so that they can meet the needs of the user. Using acoustic façade performance as an example, the user's requirement is the acoustic insulation of noise stemming from the exterior of the building. The requirement does not express values, being naturally qualitative.

The definition of the performance requirements of NBR 15575 followed the conceptual line of ISO 6147 considering the understanding that the requirements set forth in the Standard represent the requirements of users. The requirements set out below are grouped by category: Safety, Habitability and Sustainability.

- Safety
  - Structural safety
  - Fire safety
  - Safety in use and operation
- Habitability
  - Watertightness
  - Thermal performance
  - Acoustic performance
  - Lighting performance
  - Health, hygiene and air quality
  - Functionality and accessibility
  - Tactile and anthropodynamic comfort
- Sustainability
  - Durability and maintainability
  - Environmental impact

Table 2 shows the number of specific requirements for each requirement and performance category.

Given the need to present values, the standard establishes performance criteria for each requirement. By definition, criteria are quantitative specifications of performance requirements, expressed in terms of measurable quantities, so that they can be objectively determined. Following the example of acoustic performance of facades, the criterion for the requirement of acoustic insulation of external noises established by the standard is 25 decibels, considering noise class II, which represents the exposure condition.

That is, the criterion expresses a value for the performance requirement, which makes it tangible, measurable, and for the measurement of these values are established, for each requirement and performance criterion, evaluation methods. The objective of the evaluation methods is to ensure the uniformity and representativeness of the measurements and, consequently, of the analyses. That is, regardless of who is responsible for measuring the acoustic insulation of the façade, for example,

Table 2 Number of specific performance requirements contained in NBR 15575 separated by requirement and category

Category	Requirements	Nr. of specific requirements	
Safety	Structural safety	29	68
	Fire safety	20	
	Safety in use and operation	19	
Habitability	Watertightness	16	60
	Thermal performance	6	
	Acoustic performance	12	
	Lighting performance	3	
	Tactile and anthropodynamic comfort	4	
	Health, hygiene and air quality	9	
	Functionality and accessibility	10	
Sustainability	Durability and maintainability	19	21
	Environmental impact	2	
Total		149	

the result should be the same, because the method used is standardized and established in NBR 15575 itself.

In resume, the methodology of performance analysis necessarily passes through the tripod requirement, criterion and evaluation method. It is noteworthy that the requirement is an expression of the users' requirements and the exposure conditions to which the building and its parts are exposed.

Figure 2 presents schematically the methodology of performance evaluation.

Another important point defined in NBR 15575, which brings an important legal relevance to the norm, is the attribution of tasks and responsibilities of the actors

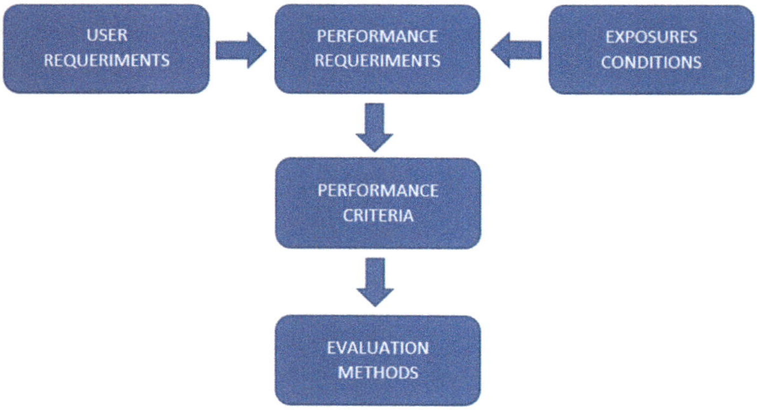

Fig. 2 Performance evaluation methodology

involved in the design, construction and use of the building. NBR 15575 classifies and assigns the following responsibilities to the actors:

- Supplier of materials, components and/or systems: It is up to the system supplier to characterize the performance, according to NBR 15575. Suppliers of components and elements shall characterize them in accordance with their applicable prescriptive standards and, if they do not exist, provide supporting results of performance based on NBR 15575 and specific international or foreign standards.
- Designer: The designer has the role of specifying materials, products and processes that meet the performance established based on prescriptive standards and the performance declared by the manufacturers of the products to be used in the project.
- Builder and developer: It is up to the developer to provide specific technical studies to identify foreseeable risks at the time of design. The builder or developer is responsible for the preparation of the manual of use and operation of the building to be delivered to the owner of the housing unit and the liquidator, in the case of the manual of use, operation and maintenance of the common areas.
- User: The user is responsible for performing maintenance in accordance with the provisions of NBR 5674 (NBR 15220-3: 2005) and the manual of use, operation and maintenance of the building.

Presenting responsibilities of the actors emphasizes the need for knowledge about the characteristics of the components and elements, including the coating layers, which make up the systems, mainly by designers and suppliers of components, elements and construction systems. "The quality of the products associated with a permanent increase in efficiency will certainly be attributes that will differentiate and privilege the companies that act with these criteria in the various sectors of the construction production chain, whose initial link, or starting point, is the design (architectural and engineering), responsible for defining the building in its fundamental characteristics (shape/design, construction system, component subsystems etc.)

1.2 Thermal Performance of Housing Buildings

NBR 15575–1 (2013) establishes that "housing building must gather characteristics that meet thermal performance requirements, considering the bioclimatic zone defined in NBR 15220-3 (NBR 5674 2012), presented in Fig. 3. This standard also presents, in its annex A, a list of the 330 Brazilian cities whose climates were classified.

In other words, the Performance Standard specifies that components, elements, and systems be used to meet thermal performance requirements and criteria. These requirements and criteria are established according to the bioclimatic zone in which the building is or will be implanted, that is, it varies according to the condition of exposure. NBR 15575-1 (2013) recommends three procedures as thermal performance

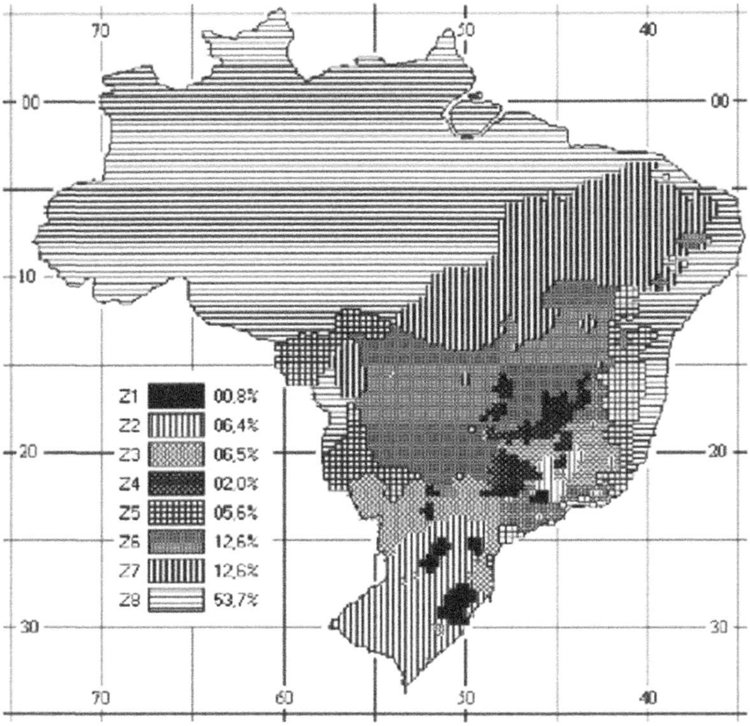

Fig. 3 Brazilian climatic zones given by NBR 15575-1 (2013)

evaluation methods: Measurement Procedure; Simplified Procedure; Computational Simulation Procedure. The measurement procedure is based on the evaluation of thermal performance via on-site temperature measurements in real-scale buildings. The measurement procedure is detailed in the Annex A of NBR 15575-1 (2013).

The difficulty of measurement on a day that is representative of a typical project day, winter or summer, brings a great uncertainty in the measurement and this method is only informative, that is, it has no normative value, evidence or proof of performance, and does not overlap with other methods: simplified or computational simulation.

The simplified procedure is based on the method of calculating the thermal properties prescribed in NBR 15220-2 (2005), more specifically transmittance and thermal capacity, and aims at the analysis of the components and elements of the building wrap, external wall systems and the roofing system.

By the definition of NBR 15575-1 (2013), thermal capacity is the amount of heat required to vary the temperature of a system in KJ/m^2 K, and thermal transmittance is the transmission of heat in a unit of time and through a unit area of an element or building component, in this case, of the glasses and opaque elements of the external walls and roofs, including the internal and external surface resistances, induced by the temperature difference between two environments.

The Performance Standard prescribes, with criteria for the simplified procedure:

- Maximum values for the thermal transmittance (U) of the external walls, as shown in Table 3;
- Minimum values for thermal capacity (TC) of the external walls, as shown in Table 4;
- Maximum values for thermal transmittance (U) of the covers, considering the descending thermal flux, according to Table 3 (Table 5).

Despite being a practical and relatively simple method of evaluating thermal performance, the simplified method in some situations presents results incompatible with reality, as in the case of external walls with large glazed areas. Simplified methods, while providing a quick tool for evaluating building performance, can understand considerable uncertainty in their results. Often these methods can compromise the process of analysis of the building in question (Sorgato et al. 2014).

If the building does not meet the requirements established by the simplified method or when the person responsible for the analysis considers that this method is inadequate for analysis, it is evaluated by the simulation method. In the simulation method, it is verified the fulfilment of the requirements and criteria established in NBR 15575-1 (2013), through computational simulation of the thermal performance of the building as a whole.

For the performance of numerical simulations and analysis of the thermal performance of the building, NBR 15575-1 (2013) recommends the use of the EnergyPlus program. This program calculates the thermal load required for heating or cooling a building environment. This calculation is based on the thermal and energetic behaviour of the building, the climate that the building is inserted, and the values of thermal loads found (Melo and Lamberts 2008). Other simulation programs can be used, provided that they allow the determination of the thermal behaviour of buildings under dynamic conditions of exposure to climate, being able to reproduce the

Table 3 Thermal transmittance of external walls

Thermal transmittance (U), W/m^2 K		
Zones 1 and 2	Zones 3, 4, 5, 6, 7 and 8	
$U \leq 2.5$	$\alpha^a \leq 0.6$	$\alpha^a \geq 0.6$
	$U \leq 3.7$	$U \geq 2.5$

[a] α is the Solar radiation absorptivity from the outer surface of the wall

Table 4 Thermal capacity of external walls

Thermal capacity (CT), KJ/m^2 K	
Zones 1, 2, 3, 4, 5, 6 and 7	Zone 8
≥130	Without requirements

Table 5 Roofs criteria for thermal transmittance

Thermal transmittance (U), W/m^2 K				
Zones 1 and 2	Zones 3 and 6		Zones 7 and 8	
U ≤ 2.3	$\alpha^a \le 0.6$	$\alpha^a \ge 0.6$	$\alpha^a \le 0.6$	$\alpha^a \ge 0.6$
	U ≤ 2.3	U ≤ 1.5	U ≤ 2.3FT	U ≤ 1.5FT

[a] α is the solar radiation absorptivity from the outer surface of the wall
Note The transmittance correction factor (FT) is established in NBR 15220-3 (NBR 5674 2012)

effects of thermal inertia and validated by ASHRAE Standard 140 (Knebel et al. 2017).

For the performance of computational simulations, NBR 15575-1 (2013) recommends the use of geographic location and climatic data corresponding to the typical summer and winter project days available in Tables A.1, A.2 and A.3 of the standard itself as a reference (NBR 15575-1 2013). However, in a statement, the Standard establishes that climate archives generated by institutions of recognized technical capacity (universities or research institutes) can be used, provided that the source is properly referenced and the data are in the public domain. For the computational model, the standard establishes that housing should be considered as a whole, considering each environment as a thermal zone, and thermal exchanges between environments. The building must be oriented according to deployment and the unit to be evaluated must be the one with the most thermally critical condition.

For the composition of the internal and external walls, floor and roof systems, the thermal properties of the components and building elements that compose it must be used. NBR 15575-1 (2013) establishes a fixed ventilation condition of an hourly air renewal (1 rph) for the environments, and the same rate for air renewal for roofing. If the building does not meet this condition, the Standard allows the increase of the ventilation rate to 5 rph. The Standard also recommends considering, for the simulation method, that the exposed walls and windows are unobstructed, that is, without the presence of buildings or vegetation nearby that modify the incidence of sun and/or wind. If the building meets this condition, it is allowed to consider shading, considering the insertion of internal or external sun capable to reduce at least 50% of the incident solar radiation into the window.

Thermal properties should be obtained through laboratory tests, following the methods presented in Table 6. In the absence of such data or the impossibility of obtaining them from manufacturers, it is permissible to use the data provided in NBR 15220-2 (NBR 15220-3 2005) as a reference.

The computational simulation method recommended in NBR 15575-1 (2013) is based on the analysis of air temperatures inside the occupy living areas, such as rooms and bedrooms, without the presence of internal heat sources (occupants, lamps, or other equipment). In addition to detailing the thermal properties of the components and elements of each layer that makes up the wall systems, as presented, in the numerical simulation is taken into account the geometric aspects of the building and the specific temperatures of the city where the location of the building. For the

Table 6 Methods of measuring thermal properties of materials and building elements

Properties	Measurement
Thermal conductivity	ASTM C518 or ASTM C177 or ISO 8302 (ASTMC 518 2017; ASTMC 177 2019; ISO 8302 1991)
Specific heat	ASTM C351-92b (1999)
Density	Measurement according to test method, preferably standardised, specific to the material
Emissivity	JIS A 1423/ASTM C1371-04[a] (2004)
Solar radiation absorptivity	ANSI/ASHRAE 74/88 (1988)
Resistance or thermal transmittance of the elements	Measurement according to NBR 6488 (1980) or calculation according to NBR 15220-2 (2005), based on thermal conductivity values measured by ASTM E903-96 (1996)
Photo-energy characteristics (glasses)	EN 410/EN 12898 (EN 410 2011; EN 12898 2019)

Table 7 Minimum criterion for evaluation of thermal performance by numerical simulation for summer conditions

Zones 1 and 8
$T_{i,máx} \leq T_{e,máx}$
$\mathbf{T_{i,máx}}$ is the maximum daily air temperature value inside the building, in °C $\mathbf{T_{e,máx}}$ is the maximum daily value of the air temperature outside the building (°C)

Note Bioclimatic zones according to the NBR 15220-3 (2012)

level of mime performance, expressed as mandatory in NBR 15575-1 (2013), and the summer condition, the maximum internal temperature cannot be higher than the maximum external temperature, as shown in Table 7, considering a typical summer day. For winter conditions, the minimum internal temperature cannot be lower than the minimum external temperature plus 3 °C, as shown in Table 8, considering a typical winter day.

Table 8 Minimum criterion for evaluation of thermal performance by numerical simulation for winter conditions

Bioclimatic zones 1–5	Bioclimatic zones 6, 7 and 8
$T_{i,mín} \geq (T_{e,mín} + 3\ °C)$	In these areas there are no criteria for winter conditions
$\mathbf{T_{i,min}}$ is the minimum daily air temperature value inside the building, in °C $\mathbf{T_{e,min}}$ is the minimum daily value of the air temperature outside the building (°C)	

Note Bioclimatic zones according to the NBR 15220-3 (NBR 5674 2012)

2 Methodology

In order to allow a comparative evaluation, a reference model environment was adopted, with typical dimensions of a room with the following internal dimensions: width equal to 2.60 m; length equal to 3.20 m; and ceiling height equal to 2.60 m.

For the analysis of thermal performance, the computational simulation method prescribed in NBR 15575-1 (2013) was used.

The reference model used in the study did not consider the existence of vegetation or buildings in the surroundings, although the influence of shading is known and relevant in the performance of the actual building. Corroborating this statement, in a research conducted in the northern hemisphere (Chan 2012) indicates that the effect caused by adjacent apartments also reduces the gain of solar heat in cold season, resulting in an increased need for energy for heating. It was considered in the façade of the model the use of sliding frames with two movable sheets with dimensions 1.20 m × 1.20 m, typical of housing buildings of economic standard. Figure 4 presents a sketch of the reference model used for numerical simulations. The environment filled in yellow was evaluated in the simulations.

2.1 Façades

For the opaque element of the external wall system (façade) of the reference model it was considered masonry of ceramic blocks of 8 holes horizontally (9 cm × 19 cm × 19 cm), internally coated with gypsum paste (thickness equal to 1 cm) and light

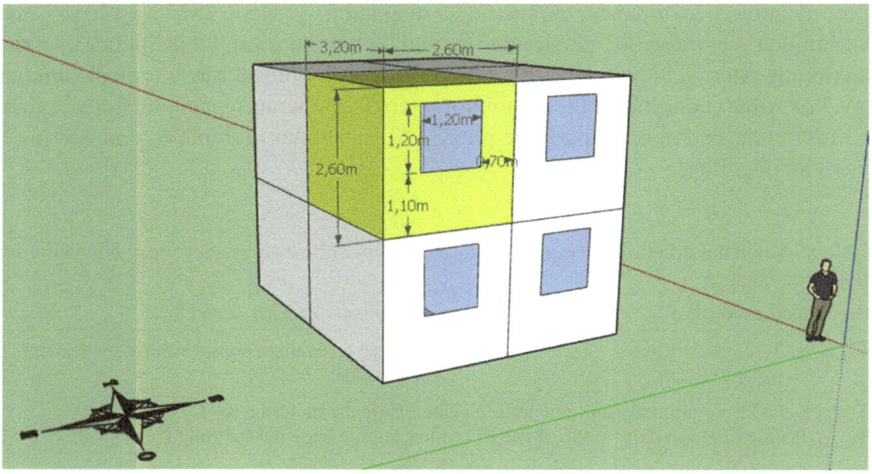

Fig. 4 Reference model used in numerical simulations

color paint and externally coated with cementitious mortar (thickness equal to 3 cm), and light colored ceramic board.

In addition to the external wall system defined for the reference model, the following variations were adopted for the elements/components with the objective of comparative analysis (see Table 9).

Table 9 External wall systems considered in numerical simulations

EWS	Internal coating		Wall	External coating		Window
1	PINα0.3	GES1	BCE9	ARG3	CERα0.3	ESQRW15FS0.85
2	*PINα0.5*	GES1	BCE9	ARG3	CERα0.3	ESQRW15FS0.85
3	*PINα0.7*	GES1	BCE9	ARG3	CERα0.3	ESQRW15FS0.85
4	PINα0.3	*GES3*	BCE9	ARG3	CERα0.3	ESQRW15FS0.85
5	PINα0.3	*ARG2*	BCE9	ARG3	CERα0.3	ESQRW15FS0.85
6	PINα0.3	*ARG4*	BCE9	ARG3	CERα0.3	ESQRW15FS0.85
7	PINα0.3	*ARG6*	BCE9	ARG3	CERα0.3	ESQRW15FS0.85
8	PINα0.3	GES1	*BCE14*	ARG3	CERα0.3	ESQRW15FS0.85
9	PINα0.3	GES1	*BCO14*	ARG3	CERα0.3	ESQRW15FS0.85
10	PINα0.3	GES1	*PCO10*	ARG3	CERα0.3	ESQRW15FS0.85
11	PINα0.3	GES1	BCE9	*ARG5*	CERα0.3	ESQRW15FS0.85
12	PINα0.3	GES1	BCE9	*ARG7*	CERα0.3	ESQRW15FS0.85
13	PINα0.3	GES1	BCE9	ARG3	*CERα0.5*	ESQRW15FS0.85
14	PINα0.3	GES1	BCE9	ARG3	*CERα0.7*	ESQRW15FS0.85
15	PINα0.3	GES1	BCE9	ARG3	*TEXα0.3*	ESQRW15FS0.85
16	PINα0.3	GES1	BCE9	ARG3	*TEXα0.5*	ESQRW15FS0.85
17	PINα0.3	GES1	BCE9	ARG3	*TEXα0.7*	ESQRW15FS0.85
18	PINα0.3	GES1	BCE9	ARG3	CERα0.3	*ESQRW19FS0.66*
19	PINα0.3	GES1	BCE9	ARG3	CERα0.3	*ESQRW23FS0.52*

Note Line 1 represents the reference model and cells filled in italics represent the layer of the wall system changed from the reference model

Legend

PINα0.3 | PINα0.5 | PINα0.7: Painting with light colour ($\alpha = 0.3$), medium ($\alpha = 0.5$) and dark ($\alpha = 0.7$)

GES1 | GES3: Gypsum plaster with thicknesses of 1 cm and 3 cm

ARG2 | ARG3 | ARG4 | ARG5 | ARG6 | ARG7: Mortar with thicknesses from 2 to 7 cm

BCE9: Masonry of 8-hole ceramic blocks horizontally 9cmx19cmx19cm

BCE14: Masonry of ceramic blocks with vertical hole 14cmx19cmx39cm

BCO14: Concrete block masonry with vertical hole 14cmx19cmx39cm

PCO10: Solid concrete wall 10 cm thick

CERα0.3 | CERα0.5 | CERα0.7: Ceramic boards with light, medium and dark colour

TEXα0.3 | TEXα0.5 | TEXα0.7: Acrylic texture with light, medium and dark colour

ESQRW15FS0.85: Window Rw = 15 dB and colourless Float glass 4 mm (Solar Factor = 0.85)

ESQRW19FS0.66: Window Rw = 19 dB and green Float glass 4 mm (Solar Factor = 0.66)

ESQRW23FS0.52: Window Rw = 23 dB and grey laminated glass 6 mm (Solar Factor = 0.52)

Table 10 Internal wall systems considered in numerical simulations

IWS	Coating 1		Wall	Coating 2	
1	PINα0.3	GES1	BCE9	GES1	PINα0.3
2	*PINα0.5*	GES1	BCE9	GES1	*PINα0.5*
3	*PINα0.7*	GES1	BCE9	GES1	*PINα0.7*
4	PINα0.3	*GES3*	BCE9	*GES3*	PINα0.3
5	PINα0.3	*ARG2*	BCE9	*ARG2*	PINα0.3
6	PINα0.3	*ARG4*	BCE9	*ARG4*	PINα0.3
7	PINα0.3	*ARG6*	BCE9	*ARG6*	PINα0.3
8	PINα0.3	GES1	*BCE14*	GES1	PINα0.3
9	PINα0.3	GES1	*BCO14*	GES1	PINα0.3
10	PINα0.3	GES1	*PCO10*	GES1	PINα0.3

Note Line 1 represents the reference model and cells filled in italics represent the layer of the wall system changed from the reference model
Legend
PINα0.3|PINα0.5|PINα0.7: Painting with light colour (α = 0.3), medium (α = 0.5) and dark (α = 0.7)
GES1|GES3: Gypsum plaster with thicknesses of 1 and 3 cm
ARG2|ARG4|ARG6: Mortar with thicknesses of 2, 4 and 6 cm
BCE9: Masonry of 8-hole ceramic blocks horizontally 9 cm × 19 cm × 19 cm
BCE14: Masonry of ceramic blocks with vertical hole 14 cm × 19 cm × 39 cm
BCO14: Concrete block masonry with vertical hole 14 cm × 19 cm × 39 cm
PCO10: Solid concrete wall 10 cm thick

2.2 Internal Wall Systems

For the internal vertical wall system of the reference model, masonry of 8-hole ceramic blocks it was considered horizontal 9 cm × 19 cm × 19 cm coated with gypsum paste (thickness ε = 1 cm) and light colour paint on both sides.

In addition to the internal wall system defined for the reference model, the following variations were adopted for the elements/components for comparative analysis (see Table 10).

2.3 Floor System

For the floor system that divides overlapping environments of the reference model was considered massive slab with 7 cm thickness, plaster board lining with thickness 2 cm and distance to the slab equal to 20 cm, without the use of mineral wool or acoustic blanket, cemented mortar floor with 3 cm thickness and coating on light coloured ceramic boards.

Table 11 Floor systems considered in numerical simulations

FS	Lining		Structure	Sub-floor		Coating
1	FOR20	S/LÃ	LAJ7	S/MAC	ARG3	CERα0.3
2	*S/FOR*	S/LÃ	LAJ7	S/MAC	ARG3	CERα0.3
3	FOR20	*LÃ5*	LAJ7	S/MAC	ARG3	CERα0.3
4	FOR20	S/LÃ	*LAJ10*	S/MAC	ARG3	CERα0.3
5	FOR20	S/LÃ	*LAJ13*	S/MAC	ARG3	CERα0.3
6	FOR20	S/LÃ	*LAJ16*	S/MAC	ARG3	CERα0.3
7	FOR20	S/LÃ	LAJ7	*MACΔ14*	ARG3	CERα0.3
8	FOR20	S/LÃ	LAJ7	*MACΔ29*	ARG3	CERα0.3
9	FOR20	S/LÃ	LAJ7	S/MAC	*ARG5*	CERα0.3
10	FOR20	S/LÃ	LAJ7	S/MAC	*ARG7*	CERα0.3
11	FOR20	S/LÃ	LAJ7	S/MAC	ARG3	*CERα0.5*
12	FOR20	S/LÃ	LAJ7	S/MAC	ARG3	*CERα0.7*

Note Line 1 represents the reference model cells filled in italics represent the layer of the wall system changed from the reference model

Legend

FOR20: Plasterboard covering with distance of 20 cm for the slab

S/FOR: Without lining

S/LÃ: No mineral wool

LÃ5: Mineral wool with thicknesses of 5 cm on the lining

LAJ7|LAJ10|LAJ13|LAJ16: Solid concrete slab with thicknesses 7, 10, 13 and 16 cm

SMAC: No acoustic blanket

MACΔ14: Acoustic blanket with ΔLw equal to 14 dB

MACΔ29: Acoustic blanket with ΔLw equal to 29 dB

ARG3|ARG5|ARG7: Cement mortar with thicknesses of 3 cm, 5 cm and 7 cm

CERα0.3|CERα0.5|CERα0.7: Ceramic boards with light colour ($\alpha = 0.3$), medium ($\alpha = 0.5$) and dark ($\alpha = 0.7$)

In addition to the floor system defined for the reference model, the following variations were adopted for the elements/components for the purpose of comparative analysis (see Table 11).

2.4 Roof System

In the roof system of the reference model it was considered massive slab with 7 cm thickness, plasterboard lining with thickness 2 cm, light colour paint and distance to the slab equal to 20 cm, without the use of mineral wool or thermal blanket, waterproofing with asphalt blanket with thickness 4 mm, mechanical protection in cementitious mortar with 5 cm thickness and coating in light colour paint.

Table 12 Roofing systems for numerical simulations

FS	Lining		Structure	Sub-floor		Coating
1	FOR20α0.3	S/LÃ	LAJ7	S/MTE	IMP0.4 + ARG5	PINα0.3
2	*FOR20α0.5*	S/LÃ	LAJ7	S/MTE	IMP0.4 + ARG5	PINα0.3
3	*FOR20α0.7*	S/LÃ	LAJ7	S/MTE	IMP0.4 + ARG5	PINα0.3
4	*S/FOR*	S/LÃ	LAJ7	S/MTE	IMP0.4 + ARG5	PINα0.3
5	FOR20α0.3	*LÃ5*	LAJ7	S/MTE	IMP0.4 + ARG5	PINα0.3
6	FOR20α0.3	S/LÃ	*LAJ10*	S/MTE	IMP0.4 + ARG5	PINα0.3
7	FOR20α0.3	S/LÃ	*LAJ13*	S/MTE	IMP0.4 + ARG5	PINα0.3
8	FOR20α0.3	S/LÃ	*LAJ16*	S/MTE	IMP0.4 + ARG5	PINα0.3
9	FOR20α0.3	S/LÃ	LAJ7	*XPS2*	IMP0.4 + ARG5	PINα0.3
10	FOR20α0.3	S/LÃ	LAJ7	*EPS4*	IMP0.4 + ARG5	PINα0.3
11	FOR20α0.3	S/LÃ	LAJ7	S/MTE	*TFC*	PINα0.3
12	FOR20α0.3	S/LÃ	LAJ7	S/MTE	IMP0.4 + ARG5	*PINα0.5*
13	FOR20α0.3	S/LÃ	LAJ7	S/MTE	IMP0.4 + ARG5	*PINα0.7*

Note Line 1 represents the reference model and cells filled in italics represent the layer of the wall system changed from the reference model

Legend

FOR20α0.3: Plasterboard lining with light colour paint (α = 0.3) and distance of 20 cm to the slab

FOR20α0.5: Plasterboard lining with medium colour paint (α = 0.7) and distance of 20 cm to the slab

FOR20α0.7: Plasterboard lining with dark colour paint (α = 0.7) and distance of 20 cm to the slab

S/FOR: No lining

S/LÃ: No mineral wood

LÃ5: Mineral wool thickness 5 cm on the lining

LAJ7|LAJ10|LAJ13|LAJ16: Solid concrete slab with thicknesses 7, 10, 13 and 16 cm

S/MTE: No thermal blanket

XPS2: Thermal blanket, XPS with 2 cm thickness

IMP0.4 + ARG5: Asphalt blanket 0.4 cm thick and mechanical mortar protection 5 cm

EPS4: Thermal blanket, EPS 4 cm thick

TFC: Roof with fibber cement tiles and tube larger than 5 cm

PINα0.3|PINα0.5|PINα0.7: Painting with light colour (α = 0.3), medium (α = 0.5) and dark (α = 0.7)

In addition to the roof system defined for the reference model, the following variations were adopted for the elements/components for the objective of comparative analysis (see Table 12).

2.5 *Thermal Performance*

For numerical simulation and thermal performance analysis of the reference model and models with variations of the defined wall systems, *Sketchup 8* and *EnergyPlus V8* software's were used. In a study to identify the trend of use of *EnergyPlus* software by researchers in Brazil (Santos et al. 2018) found that from 2004, its use began to be identified, intensifying over the years, thus indicating a trend of its greater use in future research.

Figure 5 shows that the recommendation of the use of EnergyPlus by NBR 15575-1 (2013), since its first version, published in 2010, has impacted the use of the software in academic research, as is the case of the present study. For the creation of the model considering the thermal zones and enabling the communication and transfer of data from SketchUp 8 to EnergyPlusV8, the OpenStudio plugin was used.

Figure 6 presents sketchup8 screen with the reference model inserted. At the top end of the screen you can view OpenStudio tools specific to interface with

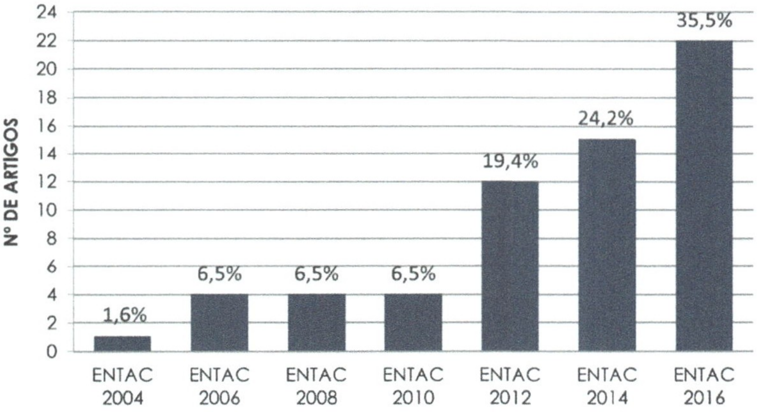

Fig. 5 Chronology of use of *EnergyPlus* in research published in ENTAC between 2004 and 2016

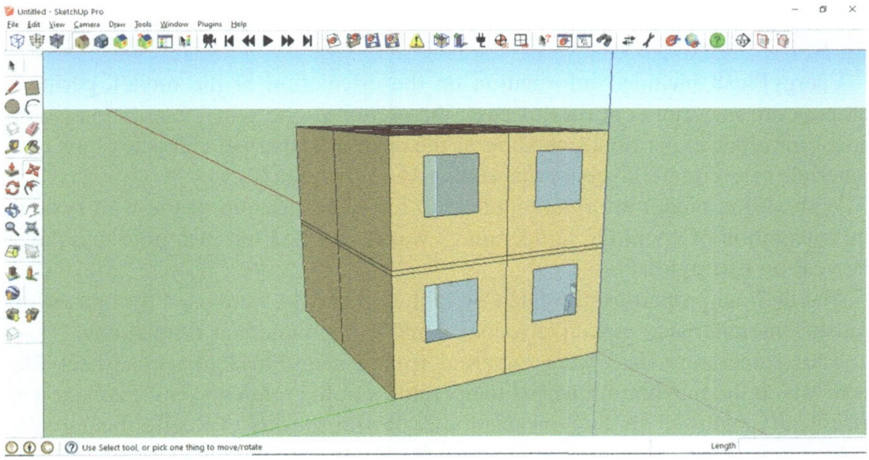

Fig. 6 Tela do SketchUp8 com o modelo de referência

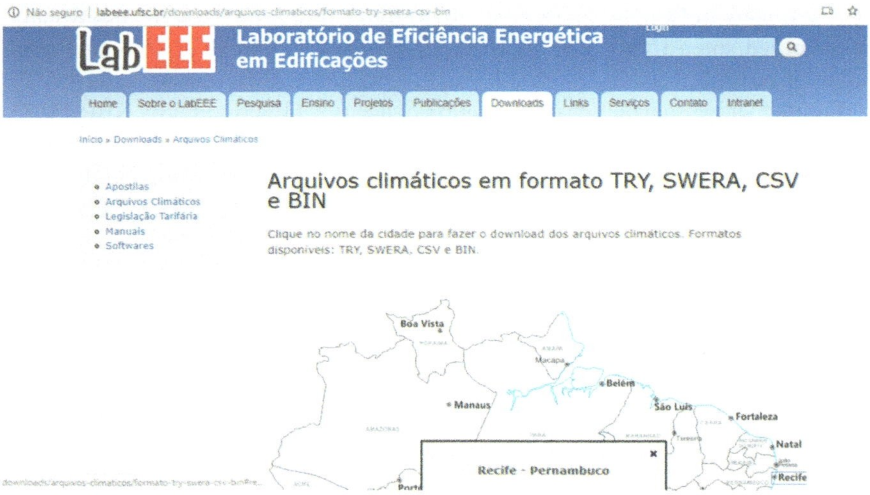

Fig. 7 Download screen of the weather files of the site labeee.ufsc.br

EnergyPlusV8. The file generated in Modelling in SketchUp8 with the OpenStudio plugin has the format ".idf" and is opened directly by EnergyPlusV8.

Climatic data from the city of Recife—PE were used. Based on a climate file in the SWERA format obtained from the website of the Laboratory of Energy Efficiency in Buildings of UFSC - Federal University of Santa Catarina (LABEEE 2019). Figure 7 presents the download page of the climate archives of the UFSC LabEEE website.

Two files were inserted on the EnergyPlusV8 home screen: the climatic file in SWERA format and the model file generated in SketchUp8, in ".idf" format, as shown in Fig. 8. The information from the sealing systems was edited directly in EnergyPlusV8 through the IDF Editor, at the direct end of the screen shown in Fig. 8.

EnergyPlus8 included information on the orientation of the model, period of temperature measurement (in this case, the period of one year), ventilation (in this case, 1air renewal per hour), shading of the openings, thermal characteristics of all materials (see Fig. 9), layers of the sealing systems (Fig. 10).

With all the model information entered, for each variation of the wall systems, the calculation of operating temperatures was performed for each hour and day of the year by EnergyPlus8.

The thermal properties presented in Tables 13 and 14 were used for opaque and translucent materials, respectively, considered in the numerical simulations.

The temperature data were exported from EnergyPlus8 to spreadsheet files, from which the maximum internal temperatures in the reference environment were specifically analysed on the typical summer day for the city of Recife, January 26.

In summary, for each variation of the wall systems, a maximum temperature value was obtained for January 26. These temperatures were analysed compared to that

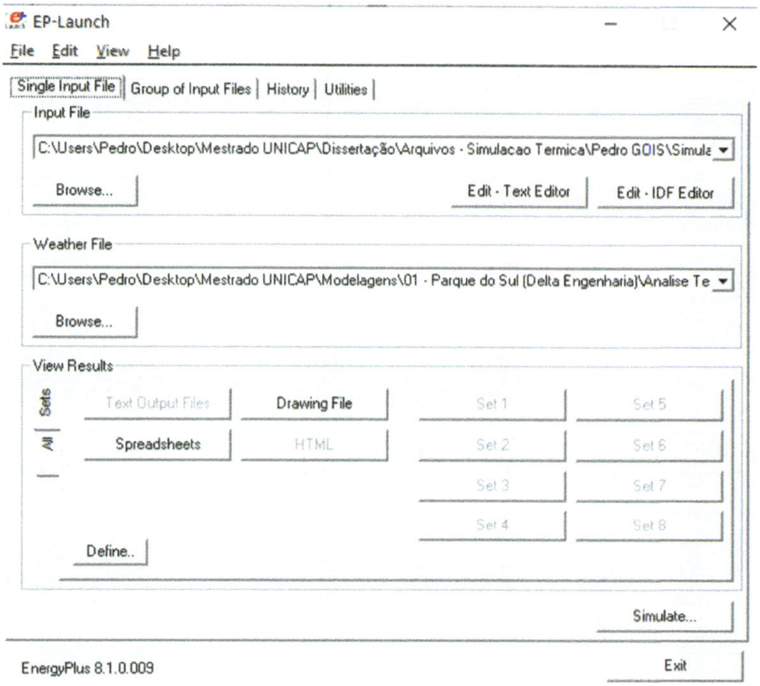

Fig. 8 Home screen of the EnergyPlusV8

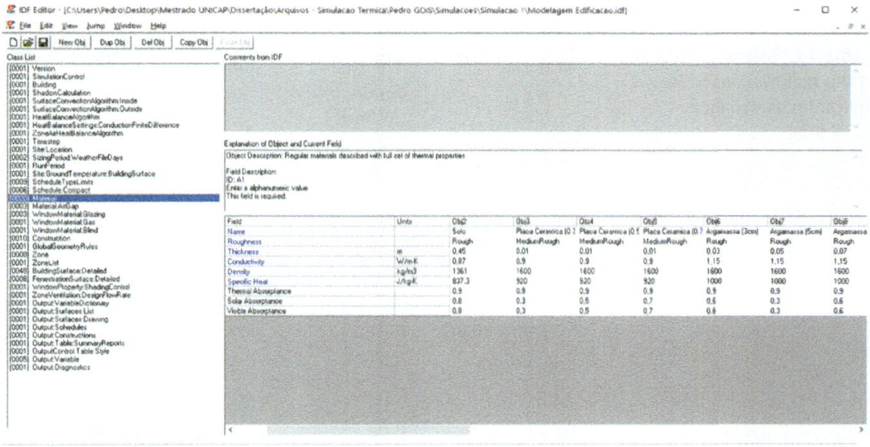

Fig. 9 "IDF" file editing screen in EnergyPlus8—Materials database

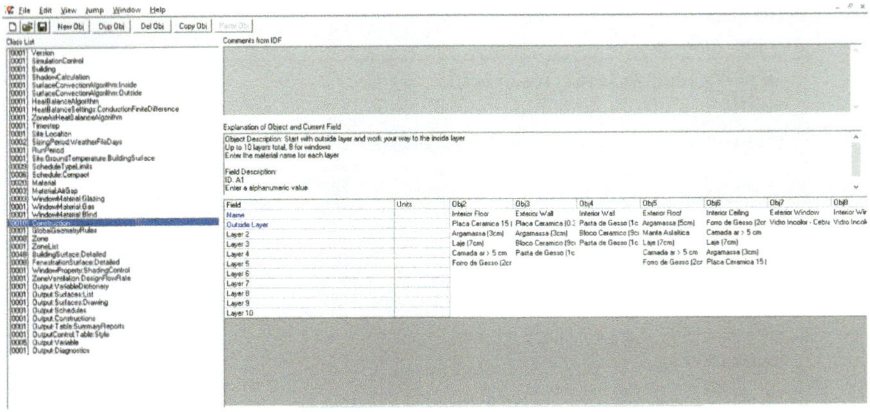

Fig. 10 "IDF" file editing screen in EnergyPlus8—Wall system layer database

Table 13 Visible radiation reflectance's used in numerical simulations

Surfaces	Reflectance bands (Dark to light colours) NBR ISO/CIE 8995-1 (2013)	Reflectance used in simulations		
		Reflectance (dark colour)	Reflectance (medium colour)	Reflectance (light colour)
Ceiling	0.6–0.9	0.6	0.75	0.9
Walls	0.3–0.8	0.3	0.55	0.8
Floor	0.1–0.5	0.1	0.3	0.5

Table 14 Thermal properties of the materials used in thermal performance modelling

Material	Thermal conductivity (W/m K)	Density (kg/m^3)	Specific heat (J/kg K)
Gypsum	0.5	1200	840
Mortar	1.15	1600	1000
Concrete	1.75	2400	1000
Ceramic block	0.7	1300	920
Ceramic board	0.9	1600	920
Gypsum board	0.35	900	840
Glass wool	0.45	22	700
XPS	0.035	40	1420
EPS	0.04	35	1420
Fibber cement tile	0.95	1900	840

Note The solar absorbance used varied according to the colour of the coating used, being considered $\alpha = 0.3$ for light colours; 0.5 for medium colours; and 0.7 for dark colours, as recommended in NBR 15575-1 (2013)

obtained in the reference model, whose results and discussions are presented in the Section below.

3 Results and Discussion

During the preliminary simulations and recording of the results obtained, it was verified that the changes made in the floor system and in the internal wall systems presented none, or very little significant variations in temperature in relation to the reference model.

This verification demonstrates that wrapping systems, external wall systems (facades) and roofing systems govern the thermal performance of the building and justify the specific analysis of these systems in the simplified evaluation method recommended by NBR 15575-1 (2013).

Thus, based on the information obtained in a preliminary analysis that the internal wall systems and floor systems had an irrelevant influence on the thermal performance of the reference model, the results and analyses are presented, considering the variations in the systems of external walls (facades) and roofing systems.

3.1 External Wall System Influence on Thermal Performance

The variations related to external wall systems (EWS) and the results obtained in the numerical simulations considering the reference model and such variations are presented in Table 15.

It is noteworthy that in these models were considered the systems of internal walls, floor systems and roofing system of the reference model according to Tables 10, 11 and 12, respectively.

Based on the results presented in Table 15, considering the model used in the study, the following analyses can be performed:

- The colour of the internal finish of the external wall does not interfere with the thermal performance of the environment (ID 47 and 48);
- The increase in the thickness of the coating layers, both internal and external, showed a direct correlation with the decrease in temperature in the evaluated environment, contributing significantly to the improvement of the thermal performance of the model (ID 49 to 52, 56 and 57). The changes in the structuring elements of the external walls also resulted in significant changes in temperature inside the environment, and the solid concrete wall presented the best result among the evaluated systems (ID 53 to 55);
- Changes in the colour of the external lining of the facades resulted and very significant variations in the temperature values measured in the environments and the darker the temperature, consequently, worse thermal performance (ID 58 to

Table 15 External walls systems considered in numerical simulations and respective thermal performance results

ID	EWS	Internal coating		Wall	External coating		Window	Temperature (°C) (increment)[a]
46	1	PINα0.3	GES1	BCE9	ARG3	CERα0.3	ESQFS0.85	33.48
47	2	*PINα0.5*	GES1	BCE9	ARG3	CERα0.3	ESQFS0.85	33.48 (0)
48	3	*PINα0.7*	GES1	BCE9	ARG3	CERα0.3	ESQFS0.85	33.48 (0)
49	4	PINα0.3	*GES3*	BCE9	ARG3	CERα0.3	ESQFS0.85	**33.32 (−0.16)**
50	5	PINα0.3	*ARG2*	BCE9	ARG3	CERα0.3	ESQFS0.85	**33.19 (−0.29)**
51	6	PINα0.3	*ARG4*	BCE9	ARG3	CERα0.3	ESQFS0.85	**32.92 (−0.56)**
52	7	PINα0.3	*ARG6*	BCE9	ARG3	CERα0.3	ESQFS0.85	**32.73 (−0.74)**
53	8	PINα0.3	GES1	*BCE14*	ARG3	CERα0.3	ESQFS0.85	**33.12 (−0.36)**
54	9	PINα0.3	GES1	*BCO14*	ARG3	CERα0.3	ESQFS0.85	**33.26 (−0.22)**
55	10	PINα0.3	GES1	*PCO10*	ARG3	CERα0.3	ESQFS0.85	**32.74 (−0.73)**
56	11	PINα0.3	GES1	BCE9	*ARG5*	CERα0.3	ESQFS0.85	**33.24 (−0.24)**
57	12	PINα0.3	GES1	BCE9	*ARG7*	CERα0.3	ESQFS0.85	**33.06 (−0.41)**
58	13	PINα0.3	GES1	BCE9	ARG3	*CERα0.5*	ESQFS0.85	***34.80 (+1.33)***
59	14	PINα0.3	GES1	BCE9	ARG3	*CERα0.7*	ESQFS0.85	***36.15 (+2.67)***
60	15	PINα0.3	GES1	BCE9	ARG3	*TEXα0.3*	ESQFS0.85	***33.61 (+0.13)***
61	16	PINα0.3	GES1	BCE9	ARG3	*TEXα0.5*	ESQFS0.85	***35.00 (+1.52)***
62	17	PINα0.3	GES1	BCE9	ARG3	*TEXα0.7*	ESQFS0.85	***36.38 (+2.90)***
63	18	PINα0.3	GES1	BCE9	ARG3	CERα0.3	*ESQFS0.66*	**33.18 (−0.30)**

(continued)

Table 15 (continued)

ID	EWS	Internal coating		Wall	External coating		Window	Temperature (°C) (increment)[a]
64	19	PINα0.3	GES1	BCE9	ARG3	CERα0.3	*ESQFS0.52*	**32.96 (−0.52)**

Note 1 Line 1 represents the reference model and cells filled in italics represent the layer of the seal system changed in relation to the reference model
Note 2 Results that show temperature increase in relation to the reference model highlighted with bold italics padding
Note 3 Results that present temperature reduction in relation to the model highlighted with bold fill
Legend
ID: Identification of numerical simulation
PINα0.3: Painting with light colour (α = 0.3)
GES1|GES3: Gypsum plaster with thicknesses of 1 and 3 cm
ARG2|ARG3|ARG4|ARG5|ARG6|ARG7: Mortar with thicknesses from 2 to 7 cm
BCE9: Masonry of 8-hole ceramic blocks horizontally 9 cm × 19 cm × 19 cm
BCE14: Masonry of ceramic blocks with vertical hole 14 cm × 19 cm × 39 cm
BCO14: Concrete block masonry with vertical hole 14 cm × 19 cm × 39 cm
PCO10: Solid concrete wall 10 cm thick
CERα0.3|CERα0.5|CERα0.7: Ceramic boards with light, medium and dark colour
TEXα0.3|TEXα0.5|TEXα0.7: Acrylic texture with light, medium and dark colour
ESQFS0.85: Window with colourless Float glass with 4 mm (Solar Factor = 0.85)
ESQFS0.66: Window with green Float glass with 4 mm (Solar Factor = 0.66)
ESQFS0.52: Window with gray laminate Float glass with 6 mm (Solar Factor = 0.52)

62). It can also be observed that models with coating on ceramic boards present temperatures slightly lower than textures with the same colour.

- It can also be observed that the lower the Solar Factor of the windows of the external frames of the model, the lower the temperature measured in the environment (ID 63 and 64).

From the results presented, it can be verified that all variations of the external wall system considered in the reference model, except for the colour of the internal finish, have significant interference in the final thermal performance.

The study presented by Mendonça et al. (2019) allowed visualizing the significance of the elements of external wall in the energy consumption of buildings.

This corroborates the results presented in the present study, demonstrating a correlation between the thermal and energetic performance of the building. Considering the results obtained in the reference model, it was verified that the main influencers of thermal performance were: the colour/absorbance of the finish of the external face of the façade; the thickness of the internal and external coatings of the façade; the typology of the structuring element of the walls and the solar factor of the glasses.

3.2 Roof System Influence on Thermal Performance

The variations related to external wall systems (EWS) and the results obtained in the numerical simulations considering the reference model and such variations are presented in Table 16.

It is noteworthy that in these models were considered the systems of external seals, internal seal systems and floor systems of the reference model according to Tables 9, 10 and 11, respectively.

Based on the results presented in Table 16, considering the model used in the study, the following analyses can be performed:

- The change in the colour of the lining, that is, the colour of the ceiling, does not significantly interfere with the temperature of the environment (ID 65 and 66);
- The removal of the plaster lining of the model represented a significant increase in the temperature measured in the environment, representing a loss of thermal performance in the building (ID 67);
- The increase in the thickness of the structural element, solid concrete slab, presented a direct correlation with the improvement of thermal performance, resulting in a relatively relevant decrease in ambient temperature (ID 69, 70 and 71);
- The use of thermal blanket, either with XPS or EPS, in the roof, even with relatively small thicknesses, proved to be an intervention with significant effect on thermal performance and reduction of ambient temperatures (ID 72 and 73);
- The roof with fiber cement tiles showed a significantly lower thermal performance than the reference model, even considering the air layer between the roof and the structural element (ID 74);
- The change in colour/absorbedness of the external finishing layer of the roofing system had a very significant impact on the thermal performance of the model, having a direct relationship between the increase in absorbedness and internal temperature (ID 75 and 76).

Considering the results obtained in the reference model, it was verified that the main influencers of thermal performance were: the colour/coating of the external face of the roof; the use of a thermal blanket; the use of lining under the structural element; increasing the thickness of the structural element; and the use of mineral wool on the lining.

3.3 Influence Level of the Simulated Variables

From the results obtained, presented in Tables 15 and 16, and considering all simulated variations, a qualitative scale was defined to classify the level of influence of the coating layers on thermal performance, presented in Table 17.

Table 16 Roofing systems (RS) considered in numerical simulations and respective thermal performance results

ID	RS	Internal coating		Structure	External coating			Temperature (°C) (increment)*
46	1	FORα0.3	S/LÃ	LAJ7	S/MTE	IMP	PINα0.3	33.48
65	*2*	*FORα0.5*	S/LÃ	LAJ7	S/MTE	IMP	PINα0.3	***33.50 (+0.02)***
66	*3*	*FORα0.7*	S/LÃ	LAJ7	S/MTE	IMP	PINα0.3	***33.49 (+0.01)***
67	*4*	*S/FOR*	S/LÃ	LAJ7	S/MTE	IMP	PINα0.3	***34.00 (+0.52)***
68	5	FORα0.3	*LÃ5*	LAJ7	S/MTE	IMP	PINα0.3	**33.21 (−0.26)**
69	6	FORα0.3	S/LÃ	*LAJ10*	S/MTE	IMP	PINα0.3	**33.26 (−0.22)**
70	7	FORα0.3	S/LÃ	*LAJ13*	S/MTE	IMP	PINα0.3	**33.10 (−0.37)**
71	8	FORα0.3	S/LÃ	*LAJ16*	S/MTE	IMP	PINα0.3	**32.99 (−0.48)**
72	9	FORα0.3	S/LÃ	LAJ7	*XPS2*	IMP	PINα0.3	**32.86 (−0.62)**
73	10	FORα0.3	S/LÃ	LAJ7	*EPS4*	IMP	PINα0.3	**32.84 (−0.64)**
74	11	FORα0.3	S/LÃ	LAJ7	S/MTE	*TFC*	PINα0.3	***33.98 (+0.50)***
75	12	FOR20α0.3	S/LÃ	LAJ7	S/MTE	IMP	PINα0.5	***34.70 (+1.15)***
76	13	FOR20α0.3	S/LÃ	LAJ7	S/MTE	IMP	PINα0.7	***35.72 (+2.25)***

Note 1 Line 1 represents the reference model and cells filled in italics represent the layer of the seal system changed in relation to the reference model
Note 2 Results that show temperature increase in relation to the reference model highlighted with bold italics padding
Note 3 Results that present temperature reduction in relation to the model highlighted with bold fill
Legend
ID: Identification of numerical simulation
FOR20α0.3: Plasterboard lining with light colour paint (α = 0.3) and distance of 20 cm to the slab
FOR20α0.5: Plasterboard lining with medium colour paint (α = 0.7) and distance of 20 cm to the slab
FOR20α0.7: Plasterboard lining with dark colour paint (α = 0.7) and distance of 20 cm to the slab
S/FOR: No lining
S/LÃ: No mineral wood
LÃ5: Mineral wool thickness 5 cm on the lining
LAJ7|LAJ10|LAJ13|LAJ16: Solid concrete slab with thicknesses 7, 10, 13 and 16 cm
S/MTE: No thermal blanket
XPS2: Thermal blanket, XPS with 2 cm thickness
IMP0.4 + ARG5: Asphalt blanket 0.4 cm thick and mechanical mortar protection 5 cm
EPS4: Thermal blanket, EPS 4 cm thick
TFC: Roof with fibber cement tiles and tube larger than 5 cm
PINα0.3|PINα0.5|PINα0.7: Painting with light colour (α = 0.3), medium (α = 0.5) and dark (α = 0.7)

Taking as reference the qualitative scale of classification in the level of influence presented in Table 17 and the numerical results presented in Tables 15 and 16, it is possible to make the classification as presented in Table 18.

Table 17 Qualitative scale for classification of the level of influence of coating layers on thermal performance

Increment (i)[a]	Level of influence
$i \leq \pm 0.5°C$	Low
$\pm 0.5°C < i \leq \pm 1.0°C$	Medium
$i > \pm 1.0°C$	High

[a]Increase relative to the result of the reference model

Table 18 Influence level of the variations adopted in thermal performance in relation to the reference model

System	System variation	Level of influence
EWS	Colour of inner coating	Low
	Thickness of the inner coating	Medium
	Wall structuring element	Medium
	Thickness of external coating	Medium
	Type and colour of external coating	High
	External frame	Medium
RS	Use of lining	Medium
	Ceiling flooring colour	Low
	Use of mineral wool on the lining	Medium
	Structural element thickness (Slab)	Medium
	Use of thermal blanket	High
	Typology of the roofing system	High
	External colour of the roofing system	High

4 Conclusions

The Brazilian standard NBR 15575 "Housing Buildings—Performance" is under review in order to be published at the end of 2020, as a new version. In accordance with this, the present work intends to help the Brazilian decision-makers and give an applied and helpful guide for designers.

Considering the responsibility of the designers to correctly specify materials to be used in the new constructions in order to meet the performance levels established in the Performance Standard—NBR 15575-1 (2013)—and by the developer, according to the Performance Profile of the Enterprise, the study presented a summary of the influence on thermal performance, for each variation tested in the reference model, as shown in Table 18.

In accordance with the numerical results and considering the reference model and methodology adopted and the main analyses, it is possible to conclude:

- Thermal performance is influenced by virtually all elements and components of external wall systems (facades) and roofing system, with the colours of the façade and roofing system, use of thermal blanket and typology of the roofing system being the main factors of influence.
- Thermal performance is not significantly influenced by the floor system or internal wall system.
- In resume, the most important ideas to the scientific community, decision-makers, engineers, and academics can be expressed in "guidelines" to improve the thermal performance:
- To improve thermal performance, several strategies can be used, from the use of external coatings of external walls and coverage with lower absorbedness, in the case of hot climates, to the use of materials of low thermal conductivity in sealing systems, such as EPS, drywall counter-wall with blanket, or coatings with low thermal conductivity. The use of glasses with smaller Solar Factors, more specifically with lower transmittance to solar radiation are also an interesting possibility.

References

ANSI/ASHRAE 55 (2004) Thermal environmental conditions for human occupancy. American Society of Heating, Refrigerating and Air-Conditioning Engineers, Atlanta, USA

ASHRAE 74/88 (1988) Method of measuring solar-optical properties of materials. ASHRAE

ASTM E903-96 (1996) Standard test method for solar absorptance, reflectance, and transmittance of materials using integrating spheres. ASTM International

ASTM C351-92b (1999) Standard test method for mean specific heat of thermal insulation. ASTM International

ASTM C1371-04[a] (2004) Standard test method for determination of emittance of materials near room temperature using portable emissometers. ASTM International

ASTM C518 (2017) Standard test method for steady-state thermal transmission properties by means of the heat flow meter apparatus. ASTM International

ASTM C177 (2019) Standard test method for steady-state thermal transmission properties by means of the heat flow meter apparatus. ASTM International

Bluyssen PM (2008) Management of the indoor environment: from a component related to an interactive top-down approach. Indoor Built Environment 17(6):483–495

Borges CAM (2008) The concept of building performance and its importance for the civil construction sector in Brazil. University of São Paulo, Brazil, São Paulo

Bsen Iso 7730 (1995) Moderate thermal environments, determination of the PMV and PPD indices and specification of the conditions for thermal comfort

Chan ALS (2012) Effect of adjacent shading on the thermal performance of residential buildings in a subtropical region. Appl Energy 92:516–522

Clausen G, Wyon DP (2008) The combined effects of many different indoor environmental factors on acceptability and office work performance. HVAC&R Res 14(1):103–113

Eduardo L, Kruger P, Zannin HT (2004) Acoustic, thermal and luminous comfort in classrooms. Build Environ 39:1055–1063

EN 410 (2011) Glass in building. Determination of luminous and solar characteristics of glazing. British Standard

EN 12898 (2019) Glass in building. Determination of the emissivity. British Standard

Goldman RF (1999) Extrapolating ASHRAE's comfort model. HVAC&R Res 5(3):189–194
Gomes JEV(2015) Building performance evaluation according to ABNT NBR 15575: adaptation to the case of rehabilitated building
Haghighat F, Donnini G (1999) Impact of psycho-social factors on perception of the indoor air environment studies in 12 office buildings. Build Environ 34:479–503
ISO 6241 (1984) Performance standards in building—principles for their preparation and factors to be considered, International Organization for Standardization, Switzerland
ISO 8302 (1991) Thermal insulation—determination of steady-state thermal resistance and related properties—guarded hot plate apparatus. ISO
Knebel DE et al (2017) Standard method of test for the evaluation of building energy analysis computer programs
Lorensi LS (2013) Critical analysis and propose advances in methodologies of the experimental performance tests to analysis based on ABNT NBR 15575 (2013). UFRS, 2015NBR 15220-3: Thermal performance in buildings Part 3: Brazilian bioclimatic zones and building guidelines for low-cost houses. Rio de Janeiro-RJ, Brazil, 2005a
Melo AP, Lamberts R (2008) The thermal balance method through computer simulation in the Energyplus program. LabEEE—Energy Efficiency Laboratory in Buildings
Mendell MJ (2003) Indices for IEQ and building-related symptoms. Indoor Air 13(4):364–368
Mendonça M et al (2019) Analysis of the influence of the external vertical fence system on the energy consumption of a residential building located in the bioclimatic zone 08. XV ENCAC 1:2120–2128
Nagano K, Horikoshi T (2005) New comfort index during combined conditions of moderate low ambient temperature and traffic noise. Energy Build 37:287–294
NBR 6488 (1980) Building components—determination of conductance and thermal transmittance—Protected hot-box method. Rio de Janeiro-RJ, Brazil
NBR 15220-2 (2005) Thermal performance in buildings Part 2: calculation methods of thermal transmittance, thermal capacity, thermal delay and solar heat factor of elements and components of buildings. Rio de Janeiro-RJ, Brazil
NBR 15220-3 (2005) Thermal performance in buildings Part 3: Brazilian bioclimatic zones and building guidelines for low-cost houses. Rio de Janeiro-RJ, Brazil
NBR 15575-1 (2013) Residential buildings—performance part 1: general requirements. Rio de Janeiro-RJ, Brazil
NBR 15575-3 (2013) Residential buildings—performance part 3: requirements for floor systems. Rio de Janeiro-RJ, Brazil
NBR 15575-4 (2013) Residential buildings—performance part 4: requirements for internal and external wall systems. Rio de Janeiro-RJ, Brazil
NBR 15575-5 (2013) Residential buildings—performance part 5: requirements for roofing systems. Rio de Janeiro-RJ, Brazil
NBR 5674 (2012) Building maintenance—requirements for maintenance management system. Rio de Janeiro-RJ, Brasil
NBR ISO/CIE 8995-1 (2013) Workplace lighting—part 1: interior. Rio de Janeiro-RJ, Brazil
Santos AC et al (2018) Use of Energyplus in Brazilian research. ENTAC 555–568:14
Sorgato MJ, Melo AP, Lamberts R (2014) Analysis of the thermal performance simulation method of the NBR 15575 standard. Cadernos de Arquitetura e Urbanismo 12:12–21

Influence of the Coating System on the Natural Lighting Performance of Buildings

A. J. Costa e Silva, P. Freitas Gois, J. M. P. Q. Delgado, A. C. Azevedo, Marcos Barbosa, and M. Gois

Abstract The influence of the coating layers on the performance of construction systems is an essential parameter to assess the use of materials and elements of the systems, in order to make the project feasible, not only economically, but also technically. Through numerical simulations based on a defined reference model for the study, the present work study the influence of different layers of floor, roof and, internal and external wall systems, on the natural lighting performance. Finally, analysing the results, for the reference model used, identified the materials and elements with the greatest influence on lighting performance—the internal finishes of the environment and the type of glass used in the external walls.

Keywords Building performance · Natural lighting performance · Coatings · Numerical simulations

---

A. J. Costa e Silva · P. F. Gois
Civil Engineering Department, Universidade Católica de Pernambuco, Recife, Brazil
e-mail: angelo.silva@unicap.br

P. F. Gois
e-mail: pedrodefg@gmail.com

J. M. P. Q. Delgado · A. C. Azevedo (✉)
Civil Engineering Department, CONSTRUCT-LFC, Universidade Do Porto, Rua Dr. Roberto Frias, s/n, 4200-465 Porto, Portugal
e-mail: antonio.costaazevedo@fe.up.pt

J. M. P. Q. Delgado
e-mail: jdelgado@fe.up.pt

M. Barbosa · M. Gois
TECOMAT Engenharia, Rua Serra da Canastra, Cordeiro, Recife, Pernambuco 39150640-310, Brazil
e-mail: marcosmb.eng@gmail.com

M. Gois
e-mail: mariama21@gmail.com

J. M. P. Q. Delgado (ed.), *Efficient and Suitable Construction*, Building Pathology and Rehabilitation 17, https://doi.org/10.1007/978-3-030-62829-1_4

1 Introduction

The word performance is widely used throughout society and has quite broad meaning. It is used for hardware evaluation, professional analysis and business and sports, for example. It is common to use performance to compare professionals and equipment and, in general, a desirable standard is defined, often informally, for comparison with the performance delivered. A more modern view of performance began to be structured in the twentieth century, where studies were proposed by the National Bureau of Sciences (NBS) during the 1920s. In the 1930s and 1940s, the first performance standards were developed and the English expression performance requirements emerged (Lorensi 2015).

After the Second World War and the consequent need to build large-scale buildings in the reconstruction movement, especially in Europe, the application of innovative construction technologies and systems at the time caused the incidence of high cases of pathological manifestations, generating high economic and social burdens. Given this scenario, the need for a more careful analysis of the performance of the construction systems used proved to be very relevant. From the end of the 1960s, the United States of America and some European countries devoted themselves to deepening their studies and striving to solidify the application of the concept of performance to buildings, the book "*Savoir batir: habitabilite, durabilite, economie des batiments*", by Gerard Blachére recognized as an important publication on the theme, which conceptualizes the performance of buildings as the behaviour in use during a given useful life (Souza et al. 2018).

For example, related to temperature and relative humidity, the thermal environment affects occupants' sensation and is considered to be the environmental factor most valorised by the occupants. Extensive studies have been conducted on thermal comfort, resulting in many thermal comfort equations. The PMV (Predicted Mean Vote) and PPD (Predicted Percentage Dissatisfied) based on Fanger's comfort equation are widely used in design guides and standards (ASHRAE 55, 2004 and ISO 7730, 1995). To establish an acceptable indoor environment, all these factors should be considered (Clausen and Wyon 2008). ASHRAE Standards 55 and 62 address different environmental factors. More recently, the equivalence of the discomfort caused by different physical qualities was examined. An equivalence of acoustic sensation to thermal sensation for short-term exposure was established. Specifically, a change in temperature of 1 °C had the same effect as a change in noise of 2.6 dB (Goldman 1999). Physical environmental parameters are all interrelated, and the feeling of comfort is a composite state involving an occupant's sensations of all these factors (Goldman 1999; Haghighat and Donnini 1999; Nagano and Horikoshi 2005; Eduardo et al. 2004).

ISO 6241 (1984) "Performance standards in building - Principles for their preparation and factors to be considered", published in 1984, is an important regulatory framework for building performance. ISO 6241 establishes general principles for the development of performance standards in civil construction expressed in functional requirements of users, linking the performance of buildings and user requirements.

Table 1 General illumination levels for natural lighting

Room	General lighting (lux) for minimum performance level M
Kitchen; Bedroom Cup/kitchen; Service area	≥ 60
Bathroom; Corridor or internal staircase; Common use corridor (buildings); Common staircase (buildings); Garages/parking lots (other environments)	Not required

Note 1 For multi-floor buildings, levels of illumination slightly below the values specified in the table above (maximum difference of 20% in any dependency) are permitted for premises located on the ground floor or on floors below the street floor
Note 2 The criteria in this table do not apply to areas that are confined or unlit
Note 3 The minimum conditions required by local law must be verified and met

The objective of this Standard was to assist ISO signatory countries in the elaboration of Performance Standards, and to serve as a guide for the selection of requirements that can be applied in each case when talking about building performance (Borges, 2008). Table 1 presents the requirements of users considered in ISO 6241 (1984).

Published in 1984, ISO 6241 (1984) is still a valid and important reference for defining the performance requirements of buildings, and perhaps its main gap in relation to contemporary requirements is sustainability, considering that at the time the theme did not have the current relevance.

Like what happened in Europe in the post-war period, in the 1970s and 1980s in Brazil, the construction of large-scale buildings induced the use of new techniques and constructive technologies. During this period, known as the "Brazilian miracle", productivity was prioritized, without clear technical criteria for evaluating the innovations adopted to enable this productivity.

Also in the 1970s and 1980s, studies on the performance of buildings were developed by IPT—Institute of Technological Research, at the request of BNH—*Banco Nacional de Habitação*, studies on the performance of buildings, considered as the first studies on the subject in Brazil.

However, the first publications on the subject in Brazil report inadequate uses of innovative construction techniques and systems related to users' functional requirements, and related to the conditions of exposures exposed to buildings. According to Ferreira, [s.d.], most of the innovative solutions implemented in the country, mainly in the construction of sets financed by the extinct BNH, had components and construction systems introduced without them having an adequate technical evaluation so that they could thus predict their behaviour during their useful life. Thus, in these cases, there was an evaluation of post-occupation performance, where users served as "guinea pigs", and they were penalized with pathological problems and maintenance and replacement costs that were a consequence of the use of new poorly developed products and without adequate technical evaluation.

In resume, the same process of development of technological innovations of post-war components and construction systems in Europe occurred in Brazil, without proper performance analysis and with a high incidence of pathological manifestations causing an economic and social burden for all actors in the construction chain.

As an example of this process of technological innovations without a more detailed performance analysis, the cases of the "coffin buildings" built in the Metropolitan Region of Recife are mentioned. Melo (2007) mentions that since 1977, landslides have been recorded in 12 buildings in Recife and 50 buildings were still banned in Olinda. The study also recalled that estimates from ITEP – *Instituto Tecnológico de Pernambuco* state that at least one in ten thousand buildings in the state can collapse. A preoccupant number of buildings could collapse, in view of the consequence of this performance failure.

According to Lorensi (2015), as a result of the events, and with the focus on promoting the quality of works and leveraging the civil construction sector, CEF – *Caixa Econômica Federal* commissioned to IPT a study on the subject, which ran from 1981 to 1997. One of the pioneering works on the subject was published in the book "Building Technology", launched by IPT in 1988, with emphasis on evaluating the performance of construction systems for housing buildings. During this period, according to Manso et al. (2007), the IPT developed, with the support of the Studies and Projects Financier (FINEP), the work entitled "Minimum Performance Standards", completed in 1995.

In 1997, *Caixa Econômica Federal* hired the IPT to review the work done in 1981, and other studies were done in the same way as that prepared in 1999 by the Brazilian Institute of Technology and Construction Quality. Considering the existence of several independently developed references, Caixa Econômica Federal and the technical environment identified the need to harmonize them, transforming them into technical standards that would further facilitate the evaluation process. For the elaboration of these Standards, Caixa Econômica Federal, with the support of Finep, financed the research project "Technical Standards for the Evaluation of Innovative Construction Systems for Housing", in 2000 (Borges 2008).

In 2000, a Study Committee and working groups were created with the objective of coordinating the discussion on the performance of buildings in the technical environment, seeking consensus for the transformation of the product into A Brazilian Standard, within the scope of ABNT. The coordinator elected to the Study Commission in 2000 was Engineer Ércio Thomaz, ipt. In a second moment, in 2004, a new coordinator was elected to the Commission for The Study of the Performance Standard Project, Carlos Alberto de Moraes Borges, who remained in office until the publication of the first version of the said Standard, on May 12, 2008, and its enforceability is expected in 2010.

In 2010, with the proximity of the expected date for effective enforceability of the Performance Standard, the construction chain understood as necessary the expansion of the discussion and a longer adaptation period so that actors in the construction chain could adapt to the requirements of the Standard. Thus, in practice, the ABNT Version NBR 15575:2008 did not come into force and had its enforceability extended to 2012. Also in 2010 was elected a new coordinator for the Study Committee, Fábio Villas Boas, who conducted a new process of discussion and revision of the text of the Standard. The new text was put into public consultation in 2012 and was published and entered into force in 2013. This version of the Standard remains in force to date, however a new study committee is open, still under the coordination of Fábio Villas Boas, conducting discussions with the aim of revising the Standard and publishing a new version in the year 2020.

1.1 Natural Lighting Performance of Housing Buildings

Visual comfort is an important factor to be considered in determining the need for lighting in a building. Built environments (internal and external) are illuminated to allow the development of visual tasks (such as reading, vision, manufactures or repairs).

Among the many elements in the indoor environment, lighting seems to have the greatest impact on the human body. Several studies have investigated the impacts of light on people from different points of view for over a century. These studies demonstrate that light has visual and non-visual influences on people. Among different lighting sources, it seems that sunlight is the most crucial and cannot be easily replaced by electric light because of its dynamic quality as well as spectral characteristics. In addition, it is the most important source of vitamin D, necessary for the strength of human bones and overall health. In addition to its role as a vitamin D production agent, natural light can improve subjective mood, attention, cognitive performance, physical activity, sleep quality and attention in students and workers. All these factors can be considered key aspects to optimize academic performance and professional performance (Shishegar and Boubekri 2016).

In order for people to develop their activities accurately, with minimal risk of accidents and with less effort and risk to visual health, there is a set of conditions that are defined as visual comfort (Sorgato et al. 2014 and Lamberts et al. 2014).

In a study conducted in two offices (Borisuit et al. 2015) concludes that different lighting conditions, particularly the availability of natural light, can be an indicator of satisfaction in the work environment. NBR 15575-1 (2013) recommends requirements for natural and artificial lighting performance, establishing that:

- During the day, the housing building facilities of the rooms, dormitories, pantry/kitchens and service areas must receive convenient natural lighting, coming from the outside or indirectly, through adjacent enclosures.
- For the night time, the artificial lighting system must provide satisfactory internal conditions for enclosure occupancy and circulation in environments with comfort and safety.

Relying solely on natural lighting, the object of this work, the Brazilian Performance Standard establishes two criteria, one for the method of evaluation by computational simulation, the other for the "on-site" measurement method, presented in Tables 1 and 2.

NBR 15575-1 (2013) establishes that simulations for evaluation of natural lighting performance are carried out considering the horizontal plane, in the morning (9:30 h)

Table 2 Daytime light factor for different housing environments

Room	General lighting (lux) for minimum performance level M
Kitchen; Bedroom Cup/kitchen; Service area	≥ 0.50%
Bathroom; Corridor or internal staircase; Common use corridor (buildings); Common staircase (buildings); Garages/parking lots (other environments)	Não requerido

Note 1 For multifloor buildings, levels of illumination slightly below the values specified in the table above (maximum difference of 20% in any dependency) are permitted for premises located on the ground floor or on floors below the street floor
Note 2 The criteria in this table do not apply to areas that are confined or unlit
Note 3 The minimum conditions required by local law must be verified and met

and afternoon (15:30 h), respectively for April 23 and October 23, and its evaluation should be carried out using the algorithm presented in NBR 15215-3 (2013), meeting the following conditions:

- Consider the latitude and longitude of the site of the work, suppose days with medium cloudiness (cloud index 50%);
- Suppose deactivated artificial lighting, without the presence of opaque obstructions (windows and open curtains, open internal doors, no clothes extended in the poles, etc.);
- Simulations for the center of the environments, at the height of 0.75 m above the level of the floor; for the case of housing estates consisting of houses or houses, consider all the typical guidelines of the different units;
- In the case of housing estates consisting of multi-floor buildings, consider, in addition to the typical orientations, the different floors and the different positions of the apartments on the floors;
- In any event, consider any shading stemming from neighbouring buildings, slopes, walls and other possible bulkheads, provided that the location and conditions of the building implementation.

It was not identified in NBR 15575-1 (2013) or in another bibliographic reference adopted in this study, justification for the days and times adopted as a reference for measuring the levels of illuminance. However, these dates and times are a reference for the homogeneity and representativeness of the evaluations through the simulation method.

Another point that NBR 15575-1 (2013) does not make very clear is the location of the measuring point in the case of integrated environments, such as kitchen/service area, living/kitchen, dormitory/closet etc.

For the "on-site" measurement evaluation method, NBR 15575-1 (2013) recommends that measurements be performed horizontally, using a portable luximeter, a maximum error of ± 5% of the measured value, in the period between 9 am and 3 pm, under the following conditions:

- measurements on days with cloud cover greater than 50%, with no precipitation;
- measurements performed with deactivated artificial lighting, without the presence of opaque obstructions (open windows and curtains, open internal doors, no clothes extended on the poles, etc.);
- measurements in the center of the environments, 0.75 m above the floor level;
- for housing estates consisting of houses and houses, consider all the typical orientations of the different units;
- in the case of housing estates consisting of multi-floor buildings, consider, in addition to the typical orientations, the different floors and the different positions of the apartments on the floors;
- at the time of the measurements there can be no incidence of direct sunlight on the luximeters under any circumstances;
- the daytime light factor (FLD) is given by the relationship between internal illuminance and external illumination in the shade, according to the following equation:

$$FLD = 100 \times \frac{Ei}{Ee}$$

where *Ei* is the enlightenment within the dependency and *Ee* is the external illumination in the shade.

The "in loco" measurement method presents a certain complexity taking into account that no methodology is established to measure the level of cloud cover in the sky. This external interference can bring important differences in the analyses, which can lead to very different results in the case of measurement on days with distinct cloudiness, for example.

Another point that the method does not make very clear is the measurement period, specified between 9 and 15 h. The solar orientation at these times is known to be opposite and, consequently, the illumination in indoor environments will be significantly different.

Moreover, like the computational simulation method, NBR 15575-1 (2013) does not make very clear, it is the location of the measurement point in the case of integrated environments, such as kitchen/service area, living room/kitchen, dormitory/closet etc. In resume, the "on-site" measurement method may present a relevant degree of uncertainty and variation between results.

2 Methodology

For simulation of natural lighting performance are used various software available in the market. In a survey conducted in the na2of the XV ENCAC, conducted in 2019, 15 articles were identified that used computational simulations of lighting

Table 3 Lighting performance simulation software used in scientific research

Software	References
TROPLUX	Teixeira et al. (2019)
DAYSIM	Garcia and Pereira (2019)
	Cristina et al. (2019)
DIALUX	Cemensati et al. (2019)
DIVA/RHINOCEROS	Gabriel et al. (2019)
	Carpanedo et al. (2019)
	Bolssoni et al. (2019)
	Fontana et al. (2019)
LICASO	Gois et al. (2019)
	Trapano (2019)
APOLUX	Techio et al. (2019)
RADIANCE/ENERGYPLUS	Rodriguez and Neves (2019)
	Plazas (2019)
AUTODESK REVIT	Figueira et al. (2019)
	Queiróz et al. (2019)

performance in their methodologies. In these 15 articles, 8 different software scans were used, presented in Table 3.

For the analysis of natural lighting performance through numerical simulation, the DIAlux Evo 8.0 software was used. Detailed of the process of insertion of the input parameters will be presented below to obtain the results of the illumination calculations of each simulated environment.

DIAlux Evo 8.0 is a free software that allows the import of the floor plan into DXF file, facilitating the modelling process. The feature was used to insert the reference model into the software as can be seen in Fig. 1.

Also on the screen shown in Fig. 1 are defined the orientation of the model (See arrow indicating the North in the lower right corner of the screen) and the coordinates referring to the location of the building (See in the lower left corner of the screen). For the case study of this work, the building located in Recife—Longitude: −34.92 | Latitude: −8.05 | time zone: −03:00—and orientation of the evaluated environment as defined in the reference model: façade with window to the west and another façade exposed to the North.

It is important to highlight that the methodology of the present research delimited the analysis to the aforementioned location and orientation of the reference model. However, the software used can be used for any location and orientation of the building object of the eventual study.

After the geometry, location and orientation were defined, the building elements and components of the reference model were inserted, including the colour of their finishes, as can be seen on the 3D display screen of DIAlux EVO 8.0, presented in Fig. 2.

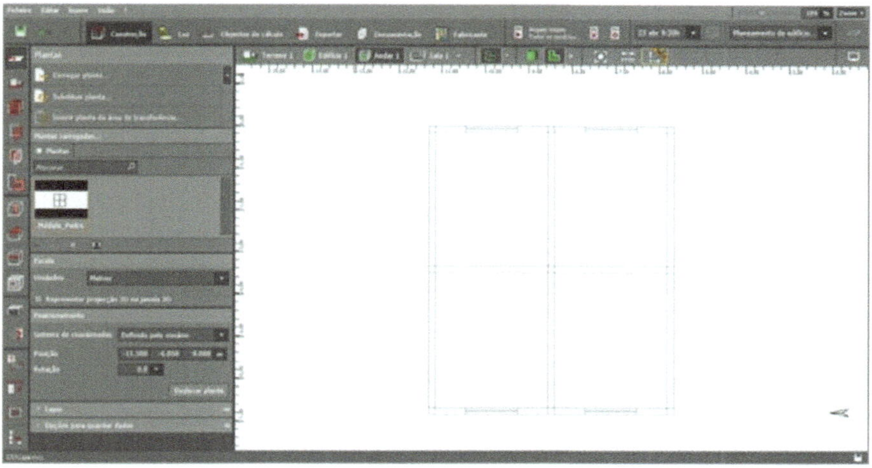

Fig. 1 DIAlux Evo 8.0 screen contemplating the floor plan of the reference model defined for the study

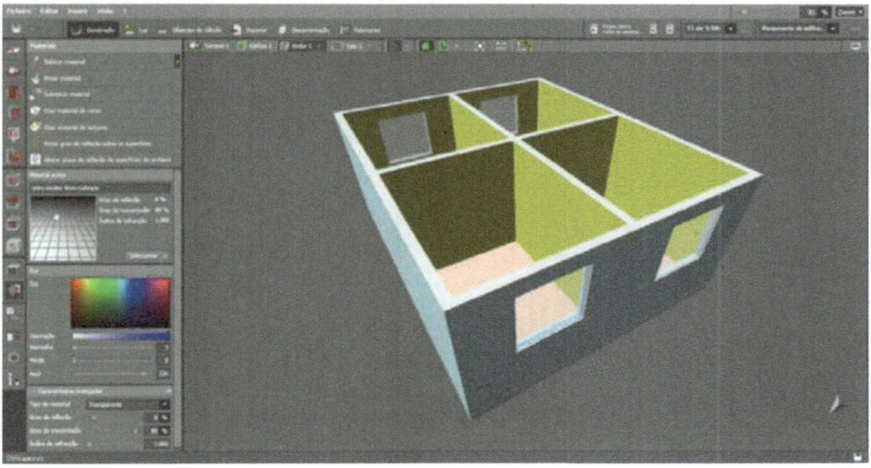

Fig. 2 DIAlux Evo 8.0 screen contemplating 3D visualization of the reference model defined for the study

The reference model defined for the study has overlapping floors, so the inclusion of the first floor was considered in DIAlux EVO 8.0, as shown in Fig. 3.

Also in Fig. 3, at its lower end, one can visualize the window of insertion of climate data, in the case of the study in question, considered the middle sky, representing a cloudiness of 50%, as recommended by ABNT NBR 15575-1 (2013).

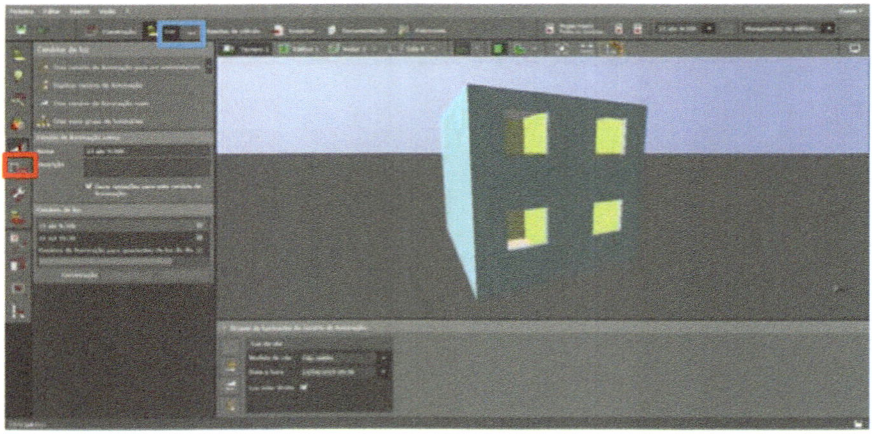

Fig. 3 DIAlux Evo 8.0 screen contemplating 3D visualization of the reference model defined for the study

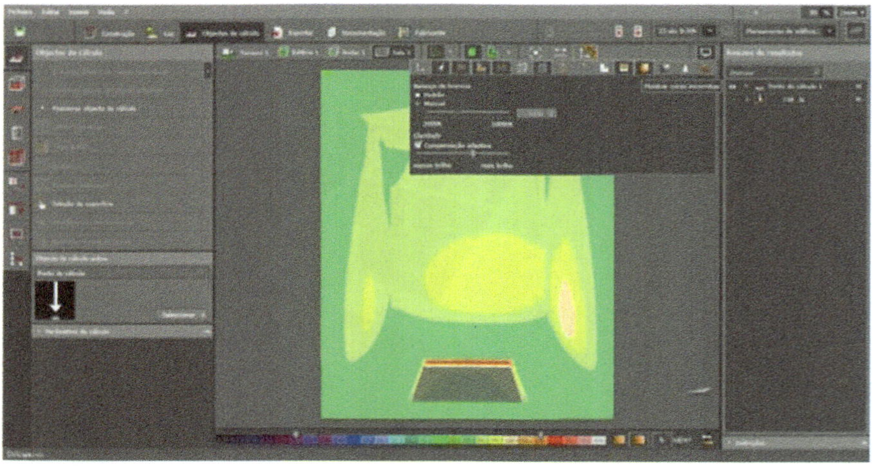

Fig. 4 DIAlux Evo 8.0 screen contemplating viewing of light destruction in the environment through color scale displayed at the bottom end of the screen

Finally, with the detailed model and all the parameters of the materials inserted in the software, the simulations were performed and the illumination levels were obtained in the center of the environment at 0.75 m high for April 23 at 9:30 am and October 23 at 3:30 pm, following the recommendation of NBR 15575-1 (2013). For the reference model and each variation of the sealing systems, illumination levels were measured for the two dates and times presented.

Figure 4 presents DIAlux screen with the feature of viewing light distribution in the environment through illumination level curves.

It is noteworthy that in the computational simulation were not considered any shading generated by the surroundings. The numerical properties presented in Tables 4 and 5 were used for coatings and glass, respectively, considered in the computational simulations.

In accordance with all defined situations modelled and the illuminance values recorded for each model, the methodology of this research consisted of the comparative analysis of the results in relation to the reference model, whose results and discussions are presented in the section below.

3 Results and Discussion

During the simulations and recording of the results obtained, it was clearly verified that the change in the thicknesses of the structural elements, walls and coating layers

Table 4 Visible reflectance's used in numerical simulations

Surfaces	Reflectance bands (Dark to light colours) NBR ISO/CIE 8995–1 (2013)	Reflectance used in simulations		
		Reflectance (Dark colour)	Reflectance (Medium colour)	Reflectance (Light colour)
Ceiling	0.6 to 0.9	0.6	0.75	0.9
Walls	0.3 to 0.8	0.3	0.55	0.8
Floor	0.1 to 0.5	0.1	0.3	0.5

Table 5 Thermal property of translucent materials (glasses) used in the modelling and numerical simulation of thermal performance

Glass	Tsol	Rsol1	Rsol2	Tvis	Rvis1	Rvis2	Emis1	Emis2
Colourless Float with 4 mm (FS = 0.85)	0.83	0.08	0.08	0.89	0.08	0.08	0.89	0.89
Green Float with 4 mm (FS = 0.66)	0.58	0.06	0.06	0.81	0.07	0.07	0.89	0.89
Grey laminate with 6 mm (FS = 0.52)	0.38	0.05	0.05	0.43	0.05	0.05	0.89	0.89

where,
Tsol = Solar transmittance (normal incidence)
Rsol1 = Solar reflectance (normal incidence) on face 1
Rsol2 = Solar reflectance (normal incidence) on face 2
Tvis = Visible transmittance (normal incidence)
Rvis1 = Visible reflectance (normal incidence) on face 1
Rvis2 = Visible reflectance (normal incidence) on face 2
Emis1 = Long-wave emissivity on the face 1
Emis2 = Long-wave emissivity on the face 2

did not generate impact at the measured illuminance level. This verification can be explained by the fact that opaque systems do not allow the passage of light, and only their surfaces generate interference in natural lighting performance.

Thus, starting from the verification performed in preliminary analysis that only the finishing layers influence the natural lighting of the environment and, consequently, in the natural lighting performance defined in the Performance Standard, will be presented in this section only the results of illuminance of the reference model and proposed variations that contemplated changes in finishing materials and translucent materials (glasses).

The variations related to the finishing layers of internal and external wall systems, floor systems and roofing systems and the results obtained in the computational simulations considering the reference model and such variations are presented in Table 6.

Based on the results presented in Table 6, considering the model used in the study, we can perform the following analyses:

- The level of illuminance in the environment has strong interference from the date and time of the check. This fact can be clearly visualized in Table 8 and justified by the orientation defined for the reference model: Window of the analysed environment facing the west in the city of Recife. It is natural that the environment is more illuminated in the late afternoon;
- The colour of the internal coating of the external wall system has a significant influence on the natural lighting performance of the model (ID 31 and 32);
- The colour of the façade lining did not present a relevant influence on the natural lighting performance of the model (ID 32 to 37);
- The types of glasses considered had a very significant influence on the natural lighting performance of the model (ID 38 and 39);
- The colour of the lining of the internal walls showed a very significant influence on the natural lighting performance of the model (ID 40 and 41);
- The colours of floor and ceiling coatings showed significant influence on the acoustic performance of the model (ID 42 to 45).

3.1 Influence Level of the Simulated Variables

From the results obtained, presented in Table 6, and considering all simulated variations, a qualitative scale was defined to classify the level of influence of the coating layers on natural lighting performance, presented in Table 7.

Taking as reference the qualitative scale of classification in the level of influence presented in Table 7 and the numerical results presented in Table 6, it is possible to make the classification as presented in Table 8.

Corroborating the result presented in Table 8, the research developed by (Husin and Harith, 2012) concludes that type of glass and window results in a major influence on the performance of natural light in the environment.

Table 6 Internal and external wall systems, floor systems and roofing systems considered in numerical simulations and their results of natural lighting performance

ID	EWS			IWS	FS	RS	23/04-09:30h		23/10-15:30h	
	Internal	External	Window				Illuminance (lux)		Illuminance (lux)	
30	PINα0.3	CERα0.3	ESQFS0.85	PINα0.3	CERα0.3	FORα0.3	738	Inc.	10938	Inc.
31	PINα0.5	CER@0.3	ESQFS0.85	PIN@0.3	CERα0.3	FORα0.3	703	-5%	10796	-1%
32	PINα0.7	CER@0.3	ESQFS0.85	PIN@0.3	CERα0.3	FORα0.3	629	-15%	10465	-4%
33	PINα0.3	CERα0.5	ESQFS0.85	PIN@0.3	CERα0.3	FORα0.3	730	-1%	10926	0%
34	PINα0.3	CERα0.7	ESQFS0.85	PIN@0.3	CERα0.3	FORα0.3	726	-2%	10919	0%
35	PINα0.3	TEXα0.3	ESQFS0.85	PIN@0.3	CERα0.3	FORα0.3	734	-1%	10932	0%
36	PINα0.3	TEXα0.5	ESQFS0.85	PIN@0.3	CERα0.3	FORα0.3	734	-1%	10932	0%
37	PINα0.3	TEXα0.7	ESQFS0.85	PIN@0.3	CERα0.3	FORα0.3	726	-2%	10919	0%
38	PINα0.3	CER@0.3	ESQFS0.66	PIN@0.3	CERα0.3	FORα0.3	651	-12%	9856	-10%
39	PINα0.3	CER@0.3	ESQFS0.52	PIN@0.3	CERα0.3	FORα0.3	348	-53%	5249	-52%
40	PINα0.3	CER@0.3	ESQFS0.85	PINα0.5	CERα0.3	FORα0.3	621	-16%	10482	-4%
41	PINα0.3	CER@0.3	ESQFS0.85	PINα0.7	CERα0.3	FORα0.3	395	-46%	9484	-13%
42	PINα0.3	CER@0.3	ESQFS0.85	PIN@0.3	CERα0.5	FORα0.3	624	-15%	10375	-5%
43	PINα0.3	CER@0.3	ESQFS0.85	PIN@0.3	CERα0.7	FORα0.3	568	-23%	10091	-8%
44	PINα0.3	CER@0.3	ESQFS0.85	PIN@0.3	CERα0.3	FORα0.5	557	-25%	10087	-8%
45	PINα0.3	CER@0.3	ESQFS0.85	PIN@0.3	CERα0.3	FORα0.7	475	-36%	9730	-11%

Legend
Inc. Percentage increase over reference model
ID Numerical simulation identification
PINα0.3 | *PINα0.5* | *PINα0.7* Painting with light colour ($\alpha = 0.3$), medium ($\alpha = 0.5$) and dark ($\alpha = 0.7$)
CERα0.3 | *CERα0.5* | *CERα0.7* Ceramic boards with light ($\alpha = 0.3$), medium ($\alpha = 0.5$) and dark colour ($\alpha = 0.7$)
ESQFS0.85 Window with colourless Float glass with 4 mm (Solar Factor = 0.85)
ESQFS0.66 Window with green Float glass with 4 mm (Solar Factor = 0.66)
ESQFS0.52 Window with gray laminate Float glass with 6 mm (Solar Factor = 0.52)
FOR20α0.3 Plasterboard lining with light colour paint ($\alpha = 0.3$) and distance of 20 cm to the slab
FOR20α0.5 Plasterboard lining with medium colour paint ($\alpha = 0.5$) and distance of 20 cm to the slab
FOR20α0.7 Plasterboard lining with dark colour paint ($\alpha = 0.7$) and distance of 20 cm to the slab

Table 7 Qualitative scale for classification of the level of influence of coating layers on natural lighting performance

Increment (i)*	Level of influence
$i \leq \pm 5\%$	Low
$\pm 5\% < i \leq \pm 20\%$	Medium
$i > \pm 20\%$	High
* *Increase relative to the result of the reference model*	

Table 8 Influence level of the variations adopted in the natural lighting performance in relation to the reference model

System	System variation	Level of influence
EWS	Colour of inner coating	Medium
	Type and colour of external coating	Low
	External frame	High
IWS	Colour of inner coating	High
FS	Colour of the floor coating	Medium
RS	Ceiling flooring color	High
	External colour of the roofing system	Low

4 Conclusions

The Brazilian standard NBR 15575 "Housing Buildings—Performance" is under review in order to be published at the end of 2020, as a new version. In accordance with this, the present work intends to help the Brazilian decision-makers and give an applied and helpful guide for designers.

Considering the responsibility of the designers to correctly specify materials to be used in the new constructions in order to meet the performance levels established in the Performance Standard –NBR 15575-1 (2013)—and by the developer, according to the Performance Profile of the Enterprise, the study presents a summary of the influence on thermal, acoustic and lighting performance for each variation tested in the reference model, as presented in Table 8.

In accordance with the numerical results and considering the reference model and methodology adopted and the main analyses, it is possible to conclude:

- The main variations that influenced the natural lighting performance were the external frames and the colour of the coatings of the internal walls and ceiling. The more translucent the glass and the lighter the colour of the walls and internal and ceiling, the higher the level of illuminance in the environment.
- The colours of the external coating and the floor also have an influence on the natural lighting performance, although with less importance.
- In resume, the most important ideas to the scientific community, decision-makers, engineers, and academics can be expressed in "guidelines" to improve the luminous performance:
- To improve the luminous performance, the use of internal coatings, especially floors and walls, with greater reflectance to visible radiation and glasses with greater transmittance to visible radiation are the best strategies for optimizing natural lighting.

References

ANSI/ASHRAE 55 (2004) Thermal environmental conditions for human occupancy. American Society of Heating, Refrigerating and Air-Conditioning Engineers, Atlanta, USA
Bolssoni G, Martinez L, Orange A, Alvarez C (2019) Analysis of the lighting performance of indoor environments from the performance of hollow elements. XV ENCAC 1(1):2639–2648
Borges CAM (2008) The concept of building performance and its importance for the civil construction sector in Brazil. University of São Paulo, Brazil, São Paulo
Borisuit A et al (2015) Effects of realistic office daylighting and electric lighting conditions on visual comfort, alertness and mood. Light Res Technol 47(2):192–209
Carpanedo FA et al (2019) Analysis of natural lighting in indoor environments of office buildings of different types in the city of Vitória-ES. XV ENCAC 1:2609–2618
Cemensati AG, Tessaro IS, Lukiantchuki MA (2019) Analysis of natural lighting in classrooms. Computer simulations and tests on Heliodon. XV ENCAC, n. 1, p. 2619–2628
Clausen G, Wyon DP (2008) The combined effects of many different indoor environmental factors on acceptability and office work performance. HVAC&R Res 14(1):103–113
Cristina K et al (2019) Lighting performance of identical windows in different cities. XV ENCAC 1:2746–2755
Eduardo L, Kruger P, Zannin HT (2004) Acoustic, thermal and luminous comfort in classrooms. Build Environ 39:1055–1063
Ferreira MA, Innovative construction systems. [s.d.]
Figueira C, Krai BDA, Oliveira MF (2019) Computational simulation of natural lighting. Analysis of the influence of the surroundings, glass and internal coatings. XV ENCAC 1:2896–2905
Fontana F, Fonseca RW, Pereira FOR (2019) Impact of the urban context on the availability of natural light in the indoor environment and on the energy consumption of artificial lighting. XV ENCAC 1:2816–2825
Gabriel E et al (2019) Comparative analysis of measurements of illuminance levels in loco and by DIVA/RHINOCEROS Plug-in simulation. XV ENCAC 1:2584–2589
Garcia DDLR, Pereira FOR (2019) Annual analysis of exposure to direct sunlight, glare and levels of natural light in an environment with internal sun protection. XV ENCAC 1:2574–2583
Gois A et al (2019) Analysis of natural and artificial lighting in the ward through computer simulations with the software AGI32 and LICASO. Study of case located at the University Hospital Gaffreé e Guinle (HUGG) - RJ. XV ENCAC 1:2599–2608
Goldman RF (1999) Extrapolating ASHRAE's comfort model. HVAC&R Res 5(3):189–194
Haghighat F, Donnini G (1999) Impact of psycho-social factors on perception of the indoor air environment studies in 12 office buildings. Build Environ 34:479–503
Husin SNFS, Harith ZYH (2012) The performance of daylight through various type of fenestration in residential building. Proc Soc Behav Sci 36:196–203
ISO 6241: Performance standards in building - Principles for their preparation and factors to be considered, International Organization for Standardization, Switzerland, 1984.
ISO 7730. Moderate thermal environments, determination of the PMV and PPD indices and specification of the conditions for thermal comfort, 1995.
Lamberts R, Dutra L, Pereira FOR (2014) Energy efficiency in architecture. Eletrobras/PROCEL
Lorensi LS (2005) Critical analysis and propose advances in methodologies of the experimental performance tests to analysis based on ABNT NBR 15575 (2013). UFRS, 2015 NBR 15220-3: Thermal performance in buildings Part 3: Brazilian bioclimatic zones and building guidelines for low-cost houses. Rio de Janeiro-RJ, Brazil
Manso MA, Mitidieri Filho CV (2007) Model of project management and coordination system for construction companies and developers in the city of São Paulo, Project Management & Technology
Melo MJAC (2007) Analysis of reports issued on "coffin-type buildings" in the metropolitan reef region: causes pointed to the landslides and interdictions, MSc Thesis, Brazil

Nagano K, Horikoshi T (2005) New comfort index during combined conditions of moderate low ambient temperature and traffic noise. Energy Build 37:287–294
NBR 15575-1: Residential buildings — Performance Part 1: General requirements. Rio de Janeiro-RJ, Brazil, 2013
NBR ISO/CIE 8995–1 - Workplace lighting - Part 1: Interior. Rio de Janeiro-RJ, Brazil, 2013
Plazas JPA (2019) Studio of sensitivity of skylights in the tropics, considering the integrated performance of luminous sufficiency, visual comfort and solar greed. XV ENCAC, p 2785–2795
Queiróz GR et al (2019) Simulation of natural lighting in Autodesk Revit according to Brazilian standards. XV ENCAC 1:2906–2915
Rodriguez LL, Neves LO (2019) Effects of the use of balconies on the luminous performance of commercial offices in office buildings. XV ENCAC 1:2776–2785
Shishegar N, Boubekri M (2016) Natural light and productivity: analyzing the impacts of daylighting on students' and workers' health and alertness. Int J Adv Chem Eng Biol Sci 3(1):1–6
Sorgato MJ, Melo AP, Lamberts R (2014) Analysis of the thermal performance simulation method of the NBR 15575 standard. Cadernos De Arquitetura E Urbanismo 12:12–21
Souza JLP, Kern AP, Tutikian BF (2018) Quantitative and qualitative analysis of the Performance Standard (NBR 15575/2013) and Main Challenges of Implementing Higher Education in Residential Multi-Floor Building. Project Manage Technol 13(1):127
Techio L et al (2019) Evaluation of natural lighting with the APOLUX program - Case study in Santa Maria. RS. XV ENCAC 1:2697–2706
Teixeira J et al (2019) The influence of the orientations of the openings and sky conditions on the levels of natural lighting in a hospital environment. XV ENCAC 1:2554–2563
Trapano PD (2019) Analysis of natural lighting levels in the classroom through measurements and simulation software AGI-32 and LICASO: Case study located in the Jorge Machado Moreira building - UFRJ. XV ENCAC 1:2649–2658

Modelling Solar Radiation and Heat Transfer of Phase Change Materials Enhanced Test Cells

A. Vaz Sá, M. Azenha, A. S. Guimarães, and J. M. P. Q. Delgado

Abstract In determining the temperature field generated within a room subjected to exterior daily variation of temperatures and solar radiation, and in view of the difficulty (or even impossibility) of obtaining analytical solutions for the descriptive differential equations of the phenomenon in most practical applications, numerical tools are used. The choice of the Finite Element Method (FEM) as a numerical methodology for solving the thermal problem associated with heat transfer in current building materials and phase change materials makes sense, as it is a well-known technique, generalize and dominated, however, still little applied to the domain of building physics. The proposed numerical simulation was performed with three main objectives: (i) the numerical resolution of the mathematical problem of heat transfer in phase change materials using the finite element method, in a three-dimensional (3D) analysis; (ii) the validation of numerical simulation capability by comparing the results obtained experimentally with the results obtained numerically; (iii) the evaluation of some of the potentialities of the numerical tool in the treatment and presentation of results. During the experimental campaign two test cells with distinct inner layers: (i) one with a reference mortar, hereinafter referred to as REFM test cell (Without PCM); (ii) another with a PCM mortar, hereinafter referred to as the PCMM test cell (With PCM). The test cells were placed outdoors and therefore have a differential effect of solar radiation. The temperatures monitored inside the REFM and PCMM test cells during the experimental campaign were compared with the values resulting from the numerical simulation using the finite element method with 3D discretization, and the obtained results revealed a significant coherence of values. An application of a solar radiation method was developed and linked, without neglecting the observation of the effect of the PCM.

Keywords Phase change materials · Finite element method · Solar radiation · Test cells · Experimental campaign

---

A. V. Sá · M. Azenha · A. S. Guimarães · J. M. P. Q. Delgado (✉)
Civil Engineering Department, Universidade do Porto, Rua Dr. Roberto Frias, s/n, 4200-465 Porto, Portugal
e-mail: jdelgado@fe.up.pt

J. M. P. Q. Delgado (ed.), *Efficient and Suitable Construction*, Building Pathology and Rehabilitation 17, https://doi.org/10.1007/978-3-030-62829-1_5

1 Introduction

Heat stored in the form of latent heat is a recognized and efficient energy storage system. The materials used for energy storage in the form of latent heat are called "Phase Change Materials" (PCM) (Lane 1983). The high storage capacity at constant temperature characterizes these materials and distinguishes them from traditional materials used in building construction.

When the environment temperature increases the chemical bonds of the PCM break and it goes from solid to liquid state. The observed chemical reaction is endothermic which means that these materials absorb the supplied heat. When the temperature drops again, the PCM returns to the solid state, releasing the stored heat. These cycles contribute for the stabilization of the interior temperature, achieved by the action of the PCM of heat conservation at constant temperature and not by increasing the thermal resistance of the buildings' elements.

Recent studies reveal the growing interest in the study of PCMs. The increase in thermal comfort inside buildings with reduced energy consumption and less gas emissions to the atmosphere remain the core drives. PCM acts by reducing inside temperature amplitudes during the day, leveling them and turning them closer to comfort temperature levels (Sá et al. 2012). Previous studies have shown that latent thermal storage is an effective way to improve energy efficiency (Halawa et al. 2005; Hasan et al. 2017; Piselli et al. 2019).

Due to the large thermal storage capacity, in the form of latent heat, PCM has been used in buildings to reduce internal temperature fluctuation (Kenisarin and Mahkamov 2007, 2016) and decrease the energy consumption of the building (Kong et al. 2017). Athienitis et al. (1997) found that the PCM plasterboard reduced the maximum ambient temperature by 4 °C. In the study presented by Kuznik and Virgone (2009), a similar result with a decrease in the maximum ambient temperature by 4.2 °C, in the cooling season, was reported for a wall composed of PCM.

The application of a cementitious mortar with PCM in small test cells revealed a reduction in the maximum temperature peak of 2.6 °C (Sá et al. 2012). The integration of PCM into cement-based building elements, such as walls and floors, has been identified as a simple and effective method for improving the energy efficiency of buildings (Cunha et al. 2016; Soares et al. 2013; Li et al. 2013). In addition, the PCM used in buildings can produce a delay in heat transfer (Lee et al. 2015). Kong et al. (2013) experimentally investigated the incorporation of PCM panels with the building wall surfaces, and found that the maximum temperature can be delayed by 2–3 h, which means a large reduction on the energy consumption for space cooling. Besides, the amount of energy savings, another important indicator for PCM applied in buildings, was also quantitatively studied in some studies. Biswas et al. (2014) have numerically studied the energy consumption of a building with PCM wallboards fabricated by nano-PCMs, indicating a reduction of more than 20% on the year-round building energy consumption. Lu et al. (2018) developed a novel system of PCM radiant floor combined with PCM wall, and obtained an average energy-saving rate of 54.27%. In order to ensure greater thermal comfort with less energy consumption,

PCM has been integrated into several building components, including walls, floors, roofs, suspended ceilings, windows, etc. (Abu-Hamdeh and Alnefaie 2019; Zhu et al. 2018). The common point of these studies is the passive use of PCM in the building.

Solar radiation presents an important role on the thermal comfort of interior spaces, particularly on passive solar buildings. However, sometimes this major parameter is neglected. The experimental campaign carried out in the present study consisted of placing test cells (with and without PCM) outdoors in order to register the effect of solar radiation. A solar model was developed and attached to the numerical resolution program used (DIANA software (Manie 2010). The temperatures monitored inside the REFM and PCMM test cells during the experimental campaign were compared with the values resulting from the numerical simulation using the finite element method with 3D discretization, and the obtained results revealed a significant coherence of values. An application of a solar radiation model was developed and linked, without neglecting the observation of the effect of the PCM.

2 Mathematical Problem

The heat transfer problems associated with the solidification and/or melting of the materials are important for the most diverse engineering areas, namely in energy storage, having been the subject of several studies.

The solution to these problems is not easy, since the interface between the solid and liquid states changes as the latent heat is stored or released. Thus, the boundary between the solid and the liquid is unknown a priori and therefore must be part of the solution (Ozisik 1993).

In this work the enthalpy method, described by Shamsundar and Sparrow (1975) and Ozisik (1993), is used as a numerical solution to the problem of heat transfer in phase-changing materials. In this method, enthalpy is temperature dependent during the change of state, in the transition from solid to liquid or from liquid to solid. Enthalpy is negative when the material stores heat and positive when the material is released, reaching the maximum value at the temperature corresponding to the melting peak and equaling zero at the beginning of the melting (corresponding to the melting temperature).

The general form of the heat equation, taking into account the variation of the enthalpy (H) with the temperature (T), is represented by (Ozisik (1993), Shamsundar and Sparrow (1975):

$$\rho \cdot \frac{\partial H(T)}{\partial t} = \lambda \cdot (\frac{\partial^2 T}{\partial x^2} + \frac{\partial^2 T}{\partial y^2} + \frac{\partial^2 T}{\partial z^2}) \tag{1}$$

Assuming linear release of the latent heat during the phase change, between T_1 and T_2 (), enthalpy variation with temperature can be assumed as (Ozisik 1993):

$$H = \{ C_p T, \text{ T } < \text{ T}_1 \quad \text{(solid)} \tag{2}$$

$$H = \left\{ C_p T + \frac{T-T_1}{T_1-T_2} L \ \mathrm{T}_1 \le \mathrm{T} \le T_2 \quad \text{(phase-change)} \right. \tag{3}$$

$$H = \left\{ C_p T + L \ \mathrm{T} > \mathrm{T}_2 \quad \text{(liquid)} \right. \tag{4}$$

where L is the latent heat and, T_1 and T_2, represent the melting temperatures (temperature at which the transition from solid to liquid begins) and peak melting temperatures (temperature at which all particles of the material are in the liquid state) - corresponding to the maximum energy storage capacity).

In regard to boundary conditions applied to the temperature field computation based on Eq. (1), the corresponding heat flows are taken into account through Eq. (5), where T is the external temperature, T_s is the surface temperature and h_{eq} is a convection/radiation coefficient that depends on air speed (Azenha 2009):

$$q = h_{eq} \cdot (T - T_s) \tag{5}$$

The effect of solar radiation on the various faces of the cells was estimated considering the real exposure conditions: air temperature and solar radiation incident on each face, using the solar radiation model described below, bringing the numerical simulation closer to the one monitored in the tests in situ.

2.1 Solar Radiation Model

The three quantities associated with the incidence of radiant energy in a body: absorbency; α reflectivity, ρ; and transmissivity, τ; are functions of the wavelength of the incident radiation. The sum of the fractions of energy absorbed, reflected and transmitted is equal to the unit.

The solar radiation absorbed by a surface, q_s, depends on the solar radiation reaching the surface, q_m, the angle of incidence, i, the absorption α_s of the surface (which is a function of color and can vary between 0 and 1).

$$q_s = \alpha_s q_m \cos(i) \tag{6}$$

where q_m represents the solar radiation reaching the surface, α_s is the absorption of the surface (which is a function of color and can vary between 0 and 1) and i is the incidence angle with the perpendicular to the surface which can be estimated as follows (Abrantes 1986):

$$\cos(i) = \cos(\alpha)\sin(h) + \sin(\alpha)s\cos(h)\cos(\psi) \tag{7}$$

This analysis of the direct solar radiation on a surface with a certain inclination implies the consideration of several geometric parameters described in Fig. 1 (α—angle that the surface makes with the horizontal plane (°), h—solar elevation (°),

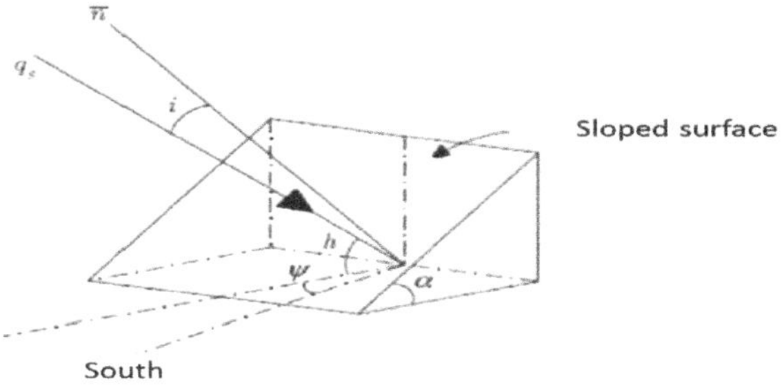

Fig. 1 Direct solar radiation on a sloped surface

corresponds to the angle between the direction of the sun's rays and the horizon, ψ – azimuth of the projection of $\overline{n}$ in the horizontal plane (°): 0° to the South, 90° to the East, 180° to the North and 270° to the West; $\overline{n}$—vector perpendicular to the surface; and i—angle of incidence of solar rays in relation to $\overline{n}$.

However, it must be emphasized that the solar model presented does not take into account the shading caused by obstacles.

3 Experimental Campaign

3.1 Pilot Test Cells

A set of experiments was conducted in order to understand the effect of the introduction of PCM into plastering mortars used as internal coatings of building spaces when exposed to solar radiation.

Two small sized closed test cells (hollow cubes with outer edge of 40 cm) were built with distinct interior coatings: (i) one with common plastering mortars, here termed as test cell REFM; (ii) another with a developed PCM mortar with formulation L, here termed as test cell PCMM. Both test cells were placed outdoors, exposed to local weather condition: solar radiation, temperature variation, relative humidity and wind speed.

The materials used for the construction of the pilot test cells were: PCMM, REFM, reinforced mortar and XPS. The main thermophysical properties of the materials used in both test cells, REFM and PCMM, are synthetized on Table 1.

The test cells were hollow cubes, with an internal hollow volume of 26 × 26 × 26 cm^3, and whose walls are composed of (from the inside to the outside of the cube): a 2 cm thick layer of mortar (REFM or PCMM); a middle layer with 3 cm thickness mortar reinforced with a steel mesh; and a 2 cm thick extruded polystyrene (XPS) layer. One of the faces of the cube functioned as a removable lid to allow access to the inside of the cell (Sá et al. 2012).

The cross-sectional composition of the walls of the test cell is not a typical one in building envelopes. In fact, the target in this case was to have a small-sized test cell,

Table 1 Thermophysical properties of the materials used in REFM and PCMM test cells (Sá et al. 2012)

Thermophysical properties	Reinforced mortar[a] (walls)	XPS[a] (external coating)	PCMM[b] (internal coating)	REFM[a] (internal coating)
Density, ρ (kg/m^3)	1500	55	1170	1400
Specific heat, c_p [kJ/(kg K)]	1.0	1.4	1.0	1.0
Volumetric specific heat, $\rho \cdot c_p$ [kJ/(m^3 K)]	1500	40	1170	1400
Thermal conductivity, λ (W/m K)	1.00	0.04	0.30	0.61
Latent heat, L (kJ/kg)	–	–	≈25	–

[a]Assumed that thermophysical properties are temperature independent, [b]Assumed no difference between liquid and solid properties

with relatively thin walls, which would however have a thermal transmittance (U ≈ 1.40 W/m^2 K) lower than the maximum limit according to Portuguese regulations for vertical elements (of U = 1.45 W/m^2 K) (Regulamento das Características de Comportamento Térmico dos Edifícios (RCCTE) 2006), thus having a reasonably similar thermal behavior to actual building envelopes. Bearing in mind that the inner plastering mortar under test should have a feasible and realistic size, a 2 cm thickness was selected (as to be able to comprise PCM within its composition). The middle layer of 3 cm steel reinforced mortar was included for structural integrity, with the thinnest possible practical width. In order to assure the desired transmittance, and bearing in mind the material characteristics [conductivity of λ = 0.04 W/m K (Building Materials and Products 2007)], the necessary thickness of polystyrene insulation was of 2 cm.

3.2 Experimental Procedure and Climatic Conditions

Both test cells were placed on the 1st floor roof of FEUP's building G, close to the LFC weather station and close to the body of the building, as shown in the Figs. 2 and 3, thus allowing the monitoring equipment to be stored inside the building and data usage of the weather station. The location where the cells were placed guarantees them a continuous daily sun exposure, during the entire period in which the tests took place: between May and June 2011.

Regarding the monitoring of temperatures, temperature sensors of the type "PT100", with a sensitivity of ±0.01 °C, were distributed inside the test cell, in a total of 10 sensors: 8 "PT100" inside (4 in each cell) and 2 "PT100" on the outside

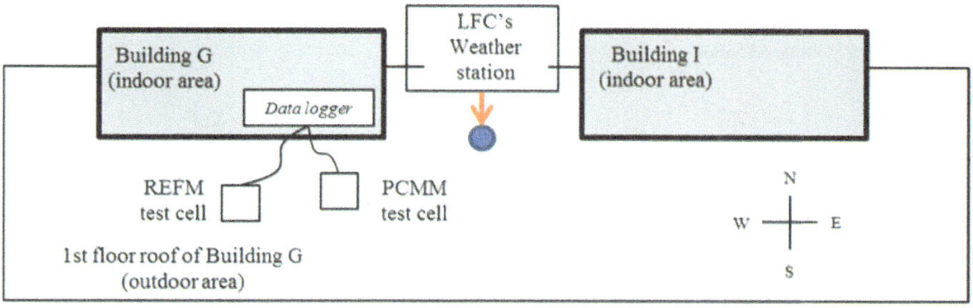

Fig. 2 Plan of FEUP's main buildings G and I: Pilot test cells outdoor arrangement set

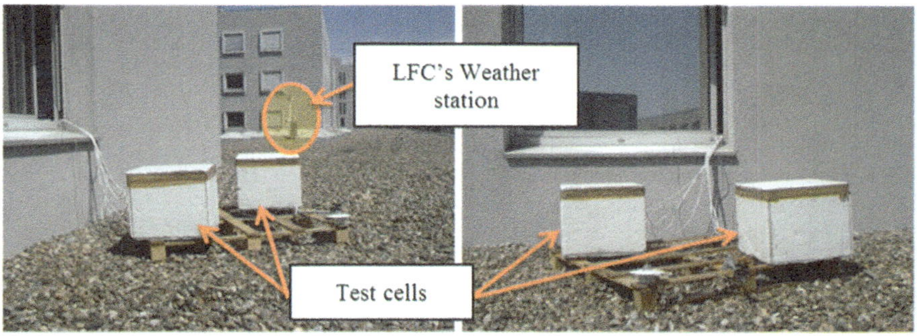

Fig. 3 In situ monitoring of test cells - Roof of the 1st floor of the building G, FEUP, Porto

(near the bottom of each test cell). The distribution of temperature sensors inside the cells was done as follows: 1 "PT100" in the center of the cover; 1 "PT100" in the geometric center; 1 "PT100" in the center of the base (bottom surface cell) and 1 "PT100" in the center of the SE face. All temperature sensors were connected to a data acquisition and storage system, with the recording of a measurement for every 10 min during the entire test cycle.

In situ experimental tests were conducted with the objective of evaluating the influence of PCM on the daily thermal performance of indoor spaces, the test cells were exposed to the climatic conditions of 4 consecutive days, between the 15th to the 19th of May 2011.

The values of the outdoor air temperature (T_{air}) and the total incident solar radiation (I_g), represented in Fig. 4, result from the acquisition of data recorded by the LFC/FEUP weather station. The ambient temperature was also recorded by two temperature sensors, type "PT100", placed on the bottom base of each of the test cells (in contact with the air and protected from solar radiation).

The first tests carried out in situ consisted of monitoring the interior temperatures of the test cells exposed to the climatic conditions represented by Figs.4 and 5. In addition to high temperatures, maximum outdoor temperature close to 30 °C, and intense solar radiation ($I_{g,max}$ = 800 W/m^2), the first two days during which

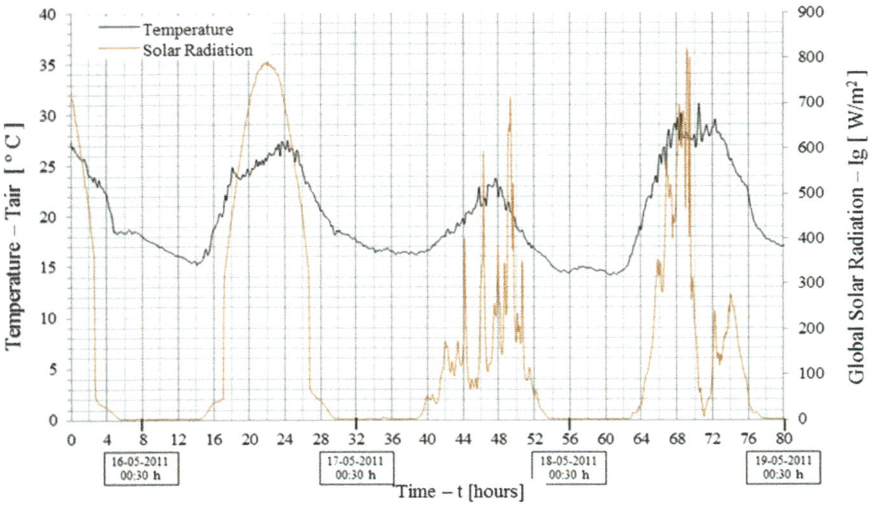

Fig. 4 Monitoring of outdoor temperature (T_{air}) and solar radiation (I_g) between 15 and 19 May 2011

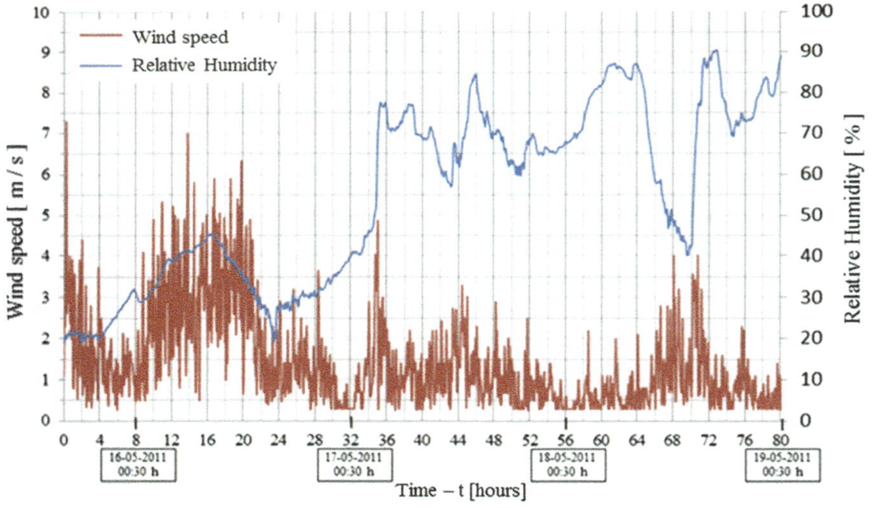

Fig. 5 Monitoring of outdoor wind speed and relative humidity between 15 and 19 May 2011

the tests took place (15 and 16 May 2011), were characterized by clear skies with very low relative humidity (RH between 19% and 45.5%), and reduced wind speed (average speed of 1.5 m/s). The next two days (17th and 18th of May) can easily be distinguished from the first by the cloudiness and the higher relative humidity (average RH of 70%).

3.3 Experimental Results and Discussion

During the first tests carried out in situ, the monitoring of the internal temperatures of the test cells shows that there is a difference between the records obtained from the sensors placed in the center of the cover and the Southeast face (SE), and the records of the sensors placed at the geometric center and at the base of the cells. This difference is observable, both in the REFM cell and in the PCMM cell, as shown in Fig. 6. The non-homogeneity in the recording of indoor temperatures, demonstrated by the recording of temperatures in different points of the interior of each of the cells, it shows the influence that solar radiation has on the exterior surface temperature and consequently on the interior temperatures (surface and ambient). Thus, the interior surfaces of the faces with greater sun exposure, such as the cell cover (horizontal top face) and the SE face, have higher temperatures than the center and base of the cell.

Taking into account the results of the monitoring of both test cells for the sensors placed on the cover and on the SE face, represented in the graph of Fig. 7, the maximum temperature recorded inside the REFM test cell was 29.2 °C, while in the PCMM test cell it was 26.0 °C. On test days with higher temperatures and more intense solar radiation, there is a difference between the maximum temperature peaks recorded inside the REFM cell and the peaks recorded inside the PCMM cell, with values of: $\Delta T = 2.9$ °C, $\Delta T = 2.6$ °C and $\Delta T = 3.2$ °C, corresponding to the 15th, 16th and 18th of May, respectively. During these days, the action of the PCM, underlined by a green spot in Fig. 5, is noticeable over several hours a day. On 16 May this action (corresponding to temperatures inside the cell in the PCM fusion

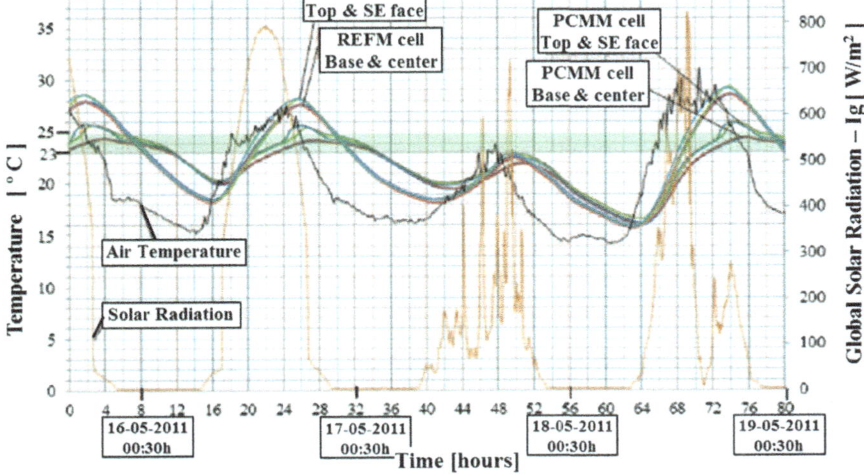

Fig. 6 Experimental results of exterior temperature, global solar radiation and internal temperature of the REFM and PCMM cells

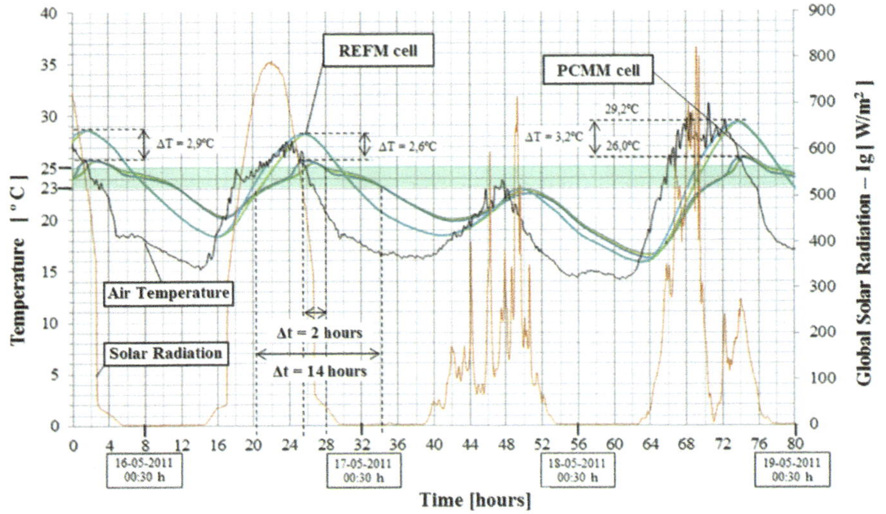

Fig. 7 Experimental results of ambient temperature, global solar radiation and temperature inside the cells (SE corner and cover)

range (from ≈23 °C to ≈25 °C) is visible for about 12 h, between 12 h and 30 min and 18 h and 30 h minutes and between 20 h and 30 min and 2 h and 30 min.

On the mildest day, corresponding to the 17th of May, the effect of the PCM is not noticeable, because the interior temperatures do not reach the melting temperature of the PCM. Throughout this day, monitoring the interior of both test cells allows us to observe that the interior temperatures (in the REFM cell and in the PCMM) are equivalent.

The monitoring carried out by the sensors placed on the base and in the geometric center of the interior of both test cells is presented in Fig. 8.

The maximum temperature recorded by these sensors inside the REFM cell was 28.8ºC, while inside the PCMM cell the maximum temperature was 24.1 °C. The differences between the maximum temperature peaks recorded inside the REFM and PCMM test cells were: ΔT = 3.9 °C; ΔT = 3.7 °C and ΔT = 4.8 °C, respectively for the 15th, 16th and 18th of May. Similar to what was observed in laboratory tests, here too there is a gap between the maximum temperature peaks recorded inside the cells. During the time that the tests were carried out, the maximum delay was approximately 2 h.

The experimental results presented by Entrop (Entrop et al. 2011) in the in situ study of small test cells consisting of a concrete floor with PCM, allowed to observe the effect of the PCM, guaranteeing, for the conditions studied, a decrease in the maximum interior temperatures and an increase minimum temperatures. In the experimental study developed, according to what was observed by Entrop et al. (2011), the action of the PCMs is felt essentially by the influence they have in the reduction of the daily peaks of temperature (maximum and minimum). The regulatory action

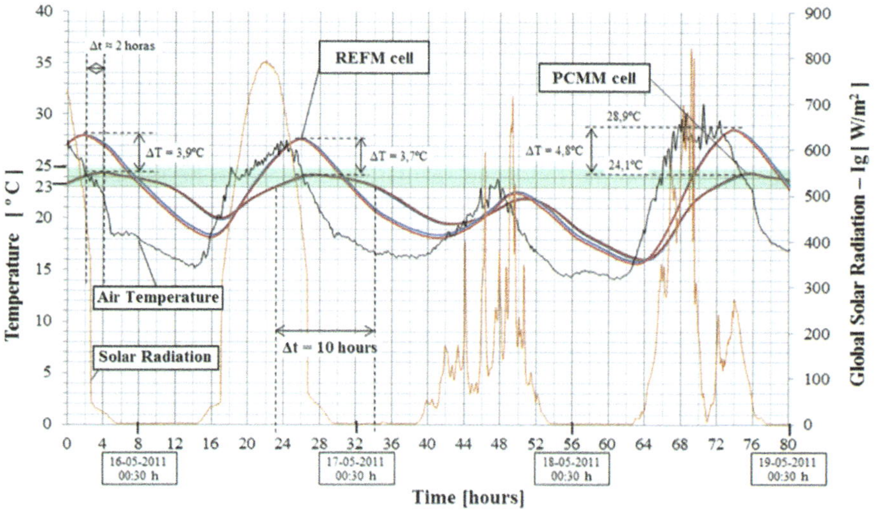

Fig. 8 Experimental results of exterior temperature, global solar radiation and internal temperature of the cells (base and geometric center)

of the temperature peaks contributes to the reduction of the temperature range registered inside and to the increase of the gap between the outside and inside temperature peaks.

Bearing in mind the collected results, it can be said that PCM acts by reducing inside temperature amplitudes during the day, levelling them and turning them closer to comfort temperature levels. It should however be stressed that the results pointed out in this research should be interpreted within its scope and not directly extrapolated to building enveloped. In fact, several issues of relevance should be taken into account: the small dimension of the test cells; and the consideration of the thermal losses only through thermal transmittance of the test cell's walls.

The main objective of the experimental investigation presented is to observe the behavior of the PCM incorporated in the developed mortar, when subjected to temperature cycles characteristic of the cooling station.

The two test cells (one coated with a standard mortar and the other coated with the mortar with PCM) were subjected in situ tests, representing scenarios corresponding to hot days in the cooling season in mainland Portugal. The monitoring of temperatures inside the two cells (REFM and PCMM) allowed the observation of the distinct behavior between them, showing the action of the PCM. PCMs contribute to the conservation of indoor temperatures between values corresponding to their melting range. Thus guaranteeing: the reduction of daily temperature peaks (maximum and minimum), the attenuation of fluctuations (or stabilization) of the indoor temperature, and the increase of the gap between the outdoor temperature peaks and the indoor temperature peaks.

The results obtained also demonstrated that the action of the PCM is not constant, being very conditioned by the conditions of exposure (of the external environment). On two consecutive days the presence of PCM can go from imperceptible to essential in conservation indoor temperature within the desired comfort limits. This fact can somehow condition the search for an optimal solution with PCM. The numerical simulation, presented in the following section, is intended to be a useful tool in the search for a solution with PCM whose performance is extended to the longest possible period of the cooling season.

4 Numerical Resolution of the Mathematical Problem

The numerical resolution of the mathematical problem of heat transfer described above is presented, performing thermal analyzes on the pilot test cells used during the experimental test campaign. The thermal analyzes developed are compared with the results obtained in the experimental monitoring in order to validate the numerical model used to solve the mathematical problem of heat transfer.

In the above-mentioned analyses, the finite element method (FEM) was used to support thermal modelling, considering a three-dimensional (3D) analysis of the heat transfer problem.

The mathematical formulations mentioned above are implemented in the DIANA software (Manie 2010) which was used for the present research work. 3D thermal fields were simulated with recourse to 8-node brick elements ($2 \times 2 \times 2$ Gauss integration) for mortar/XPS/air and 4-node planar elements (2×2 Gauss integration) for convection/radiation boundaries.

4.1 *Model Geometry and Parameters*

In modeling the complete test cell, in its true size, the following external dimensions were considered: $0.40 \times 0.40 \times 0.40$ m^3. A value of $h_{eq} = 15$ W/(m °C) was considered for the convection coefficient (Lane 1983), taking into account the exposure conditions: area limited by the presence of several buildings and reduced wind speed felt during the period in which the experimental campaign was carried out. The initial temperature inside the cells was fixed at 25 °C, similar to the temperature recorded inside the cells when the in situ monitoring started. Thermal analyzes were carried out covering a period corresponding to the days monitored during the experimental campaign, approximately 4 days, divided into 10-min increments.

The evolution of temperatures over the simulated time corresponds to the monitored temperatures during the experimental campaign by the LFC weather station. The solar radiation used in the numerical simulation took into account the values obtained through the model described above. Through this model, the values of solar radiation incident on a horizontal surface (from $\alpha = 1$), between 15 and 19 May

2011, were estimated. The approximation of the simulated values to the monitored values was achieved by adjusting the turbulence value, through an iterative trial and error process. The turbulence obtained, which best approximated the simulated radiation values to those monitored by the weather station's pyrometer, on clear skies (15 and 16 May 2011), was T = 4.5. The remaining parameters considered were $\alpha = 0°$ (horizontal surface) and latitude = 41.178°, the value of ψ is irrelevant for horizontal surfaces. The record of the simulated values and the values measured by the weather station's pyrometer for solar radiation on a horizontal surface (with absorption coefficient $\alpha = 1$) are shown in Fig. 7. Note that there is a delay in relation to the solar noon which, for the place and the days of the year under analysis, corresponds to 1 h and 34 min (Aguiar 2020), an interval already corrected in Fig. 9.

On May 15th and 16th, clear days, the simulated values are practically coincident with the values monitored by the pyrometer, following the radiation curve resulting from the monitored data, both in terms of radiation intensities and in terms of amplitude (or hours of sunshine). In the remaining days, 17th and 18th of May, the deviations between the simulated values and the measured values are justified by intermittent shading, caused by the passage of clouds.

Taking into account the proximity between the simulated and monitored values for solar radiation incident on a horizontal surface, the radiation absorbed by each of the 5 exposed surfaces of the test cells was estimated. A value of $\alpha s = 0.25$ was considered as the solar absorption coefficient of the cell's outer covering (XPS, white) (Freitas et al. 2003). In the simulation of the 4 vertical surfaces, a value of $\alpha = 9°$ was considered with values of ψ different according to their solar orientation

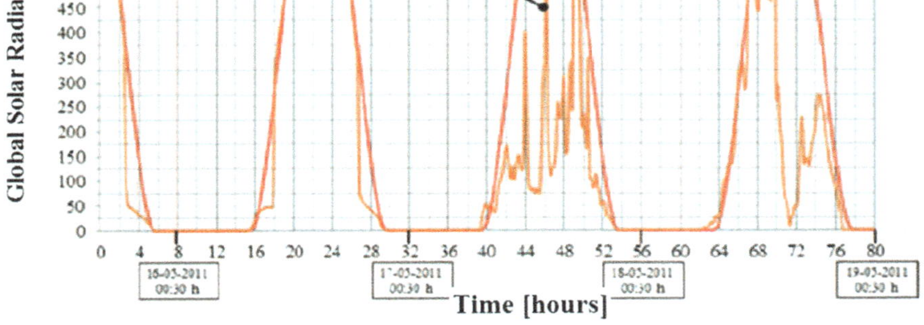

Fig. 9 Solar radiation monitored by the pyranometer and estimated by the solar model, for $T_l = 4.5$, between 15 and 19 May

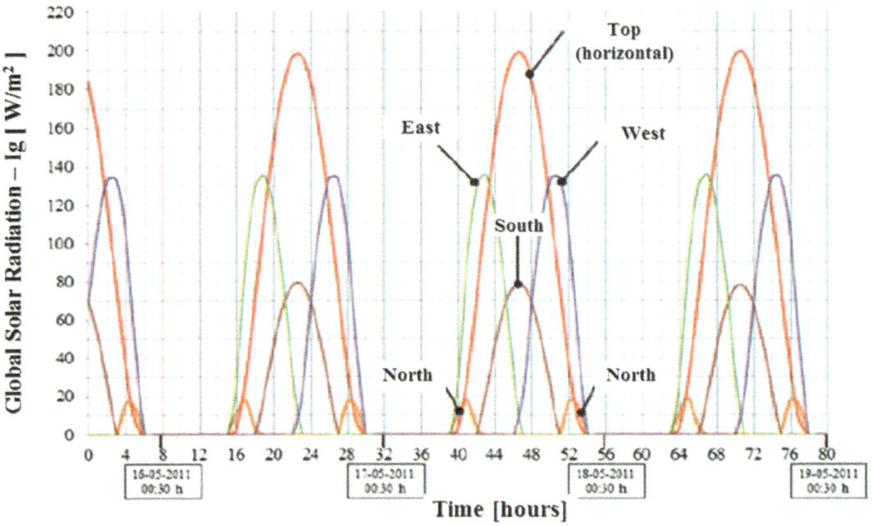

Fig. 10 Absorbed radiation, by a white surface ($\alpha s = 0.25$), estimated by the solar model for $T_l = 4.5$, between 15 and 19 May

of each face: ΨNorth = 180°; ΨSouth = 0° ΨEste = 90° and ΨWest = 270°. This simulation led to the results shown in Fig. 10.

Bearing in mind the influence of incident solar radiation on the temperatures recorded in the various points inside the test cells, it was felt necessary to bring the radiation values monitored on the 17th and 18th (with clouds) closer to the simulated values. The relationship between the monitored values and the values calculated by the model for a horizontal surface (with $\alpha = 1$) was found, that is, the relationship between the values represented in Fig. 7. Thus, the solar radiation calculated by the model and subsequently corrected (affected by the parameter that relates the monitored radiation to the radiation calculated for clear days), was estimated, having represented the values obtained in the graphs of Figs. 11 and 12.

During the numerical simulation, the two described situations were studied. Considering for the four days under analysis both the results given by the solar radiation model and the affected results of the correction.

4.2 Simulation Results and Comparison with Monitored Temperatures

The results represented in the following figures (Figs. 11, 12 and 13) reflect the values obtained in the numerical simulation for the inside of the test cells (PCMM and REFM) when affected by the ambient temperature (experimentally monitored)

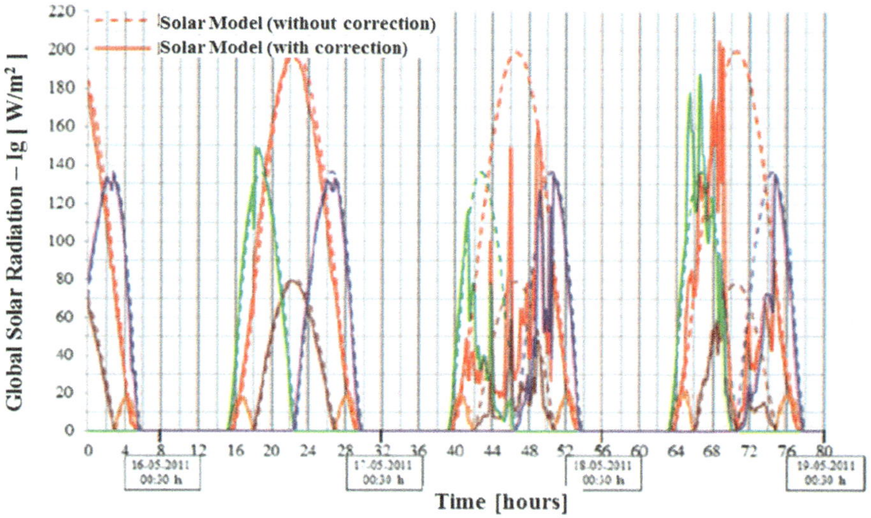

Fig. 11 Absorbed radiation, by a white surface ($\alpha s = 0.25$), estimated by the solar model for $T_l =$ 4.5 and corrected for the observed cloudiness, between 15 and 16th of May

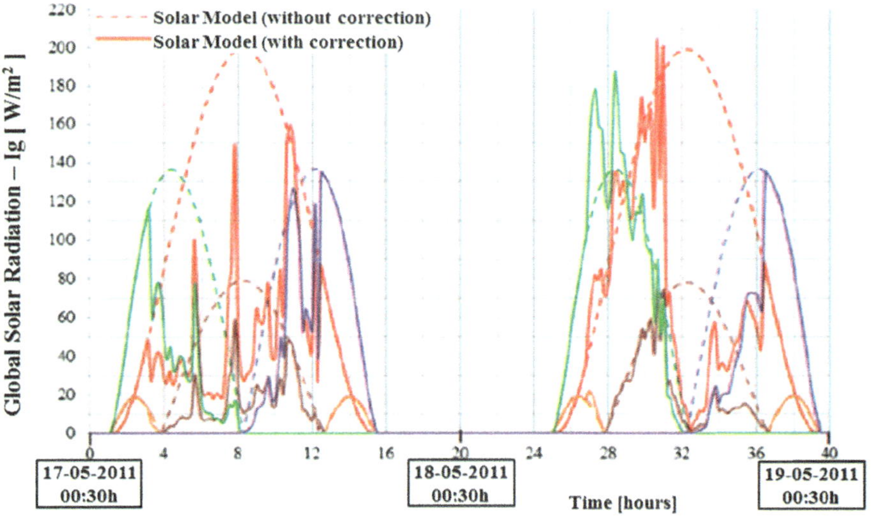

Fig. 12 Absorbed radiation, by a white surface ($\alpha s = 0.25$), estimated by the solar model for $T_l =$ 4.5 and corrected for the observed cloudiness, between 17 and 18th May

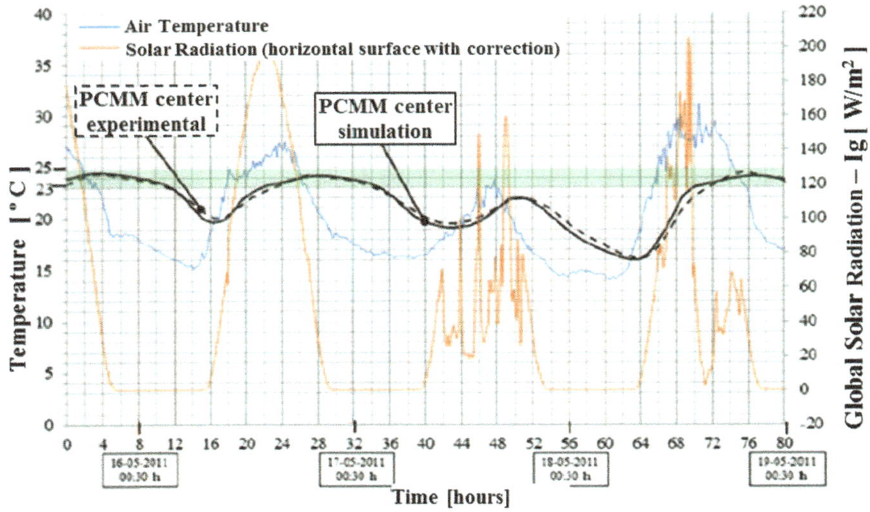

Fig. 13 In situ monitoring and numerical simulation of the temperatures in the center of PCMM test cell

and by the values of the solar radiation absorbed by each one of the 5 exposed faces of the cells calculated using the solar model adopted.

The values obtained by the numerical simulation using the solar model are, for clear days, coincident with the values monitored experimentally. Thus, for the 15th and 16th of May, clear days, the numerical simulation translates, for each point inside the cells, the observed during the experimental campaign. On cloudy days, 17 and 18 May, the simulated values show slightly higher temperatures, with a variation from the monitored values that reaches $\Delta T \approx 2$ °C in the REFM test cell and $\Delta T \approx 1$ °C in the PCMM test cell. This difference was to be expected since the solar radiation values considered in the numerical simulation are higher than those actually observed on cloudy days (17th and 18th of May).

In an attempt to approximate the conditions admitted in the simulation to the real conditions of in situ exposure, the radiation estimated by the solar model of a parameter was able to translate the cloudiness recorded over the monitored days. Thus, as shown in Fig. 9, the solar radiation of the first days (15 and 16 May) is practically unaffected, since they are clear days, since the remaining monitored days are affected by the correction parameter which is in average of 0.3 for the 17th of May and 0.6 for the 18th of May.

The results obtained (recorded in Figs. 13, 14, 15 and 16) reflect the approximation made to the actual exposure conditions. Thus, both in the REFM test cell and in the PCMM test cell, the simulated values for each point inside the cells approximate the values monitored during the 4 days of the experimental campaign, both for clear days and for clear days, the cloudy sky days.

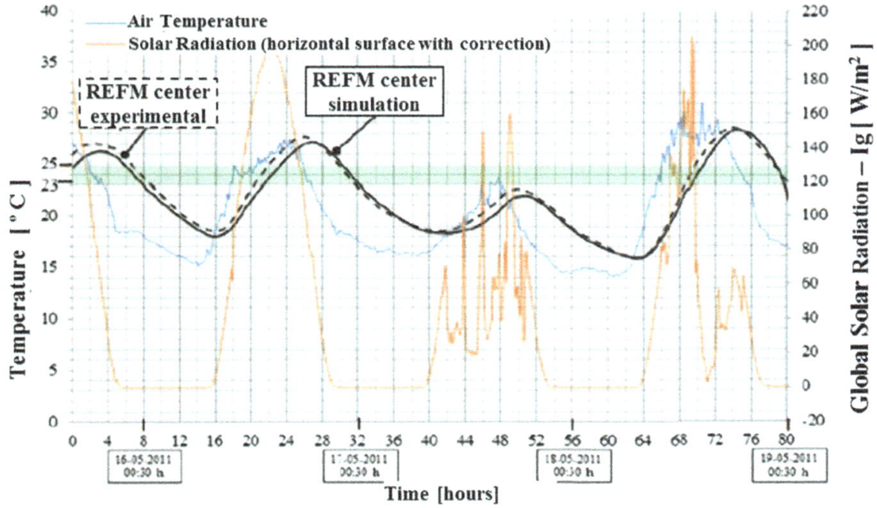

Fig. 14 In situ monitoring and numerical simulation of the temperatures in the center of REFM test cell, between 15 and 19th May

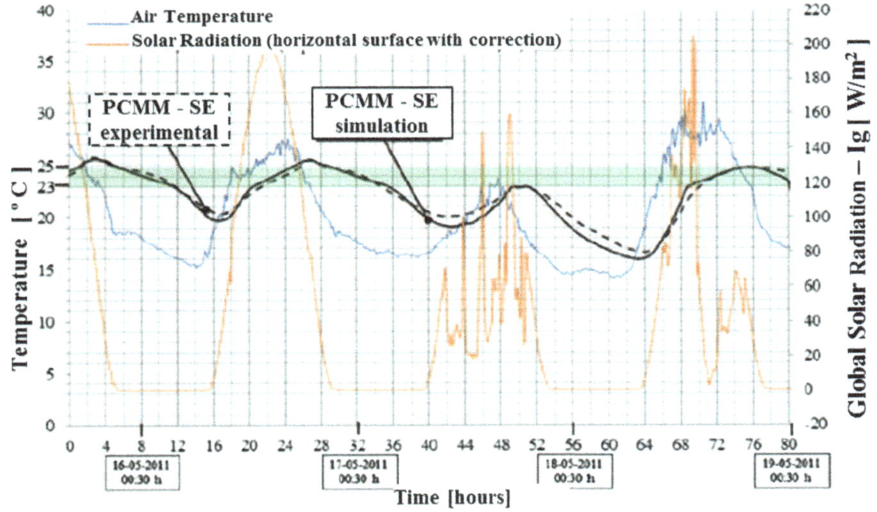

Fig. 15 In situ monitoring and numerical simulation of the temperatures in the cover of PCMM test cell

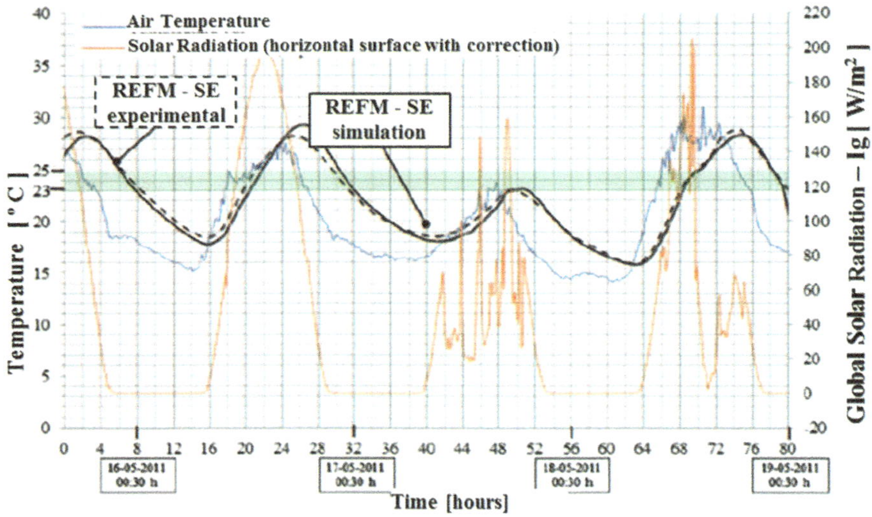

Fig. 16 In situ monitoring and numerical simulation of the temperatures in the cover of REFM test cell, between 15 and 19th May

In summary, the coupling of the solar radiation model served to bring the simulation closer to the real exposure conditions of the cells monitored in situ. It was possible to observe, in the numerical simulation, the effect of solar radiation on indoor temperatures without discouraging the main objective of the simulation performed: to observe the different behavior between REFM and PCMM cells, that is, to observe the behavior of the materials of phase change.

5 Conclusions

The numerical resolution of the mathematical problem of heat transfer in PCM was validated by the similarity between the results of the numerical simulation and the results of the experimental monitoring.

The use of the solar radiation model allowed simulate the effect of incident solar radiation on each face of the test cell (with different solar exposures) without neglecting the main objective of the recommended numerical simulation: the study of the action of PCM.

It was evident, in the numerical resolution of the heat transfer problem, similarly to that recorded experimentally, that the PCMs contribute to the conservation of indoor temperatures, ensuring a smaller fluctuation of indoor temperature, reducing peak indoor temperatures (maximum and minimum); and increasing the mismatch between indoor and outdoor temperatures.

The temperatures monitored inside the REFM and PCMM test cells during the experimental campaign Sect. 4 were compared with the values resulting from the numerical simulation, using the finite element method using 3D discretization, and the results obtained revealed a significant coherence of values.

It was possible to apply the solar radiation method, having approached the numerical resolution of the experimental tests carried out in situ, without neglecting the observation of the PCM effect.

References

Abrantes V (1986) Análise numérica e experimental do comportamento térmico de coberturas com desvão. Mestrado Dissertação de Mestrado, Faculdade de Engenharia, Universidade do Porto, Porto

Abu-Hamdeh NH, Alnefaie KA (2019) Assessment of thermal performance of PCM in latent heat storage system for different applications. Solar Energy 177:317–323

Aguiar R (2020) INETI Synthetic weather data for Portugal [Online]. Available: https://apps1.eere.energy.gov/buildings/energyplus/cfm/weather_data3.cfm/region=6_europe_wmo_region_6/country=PRT/cname=Portugal

Athienitis AK, Liu C, Hawes D, Banu D, Feldman D (1997) Investigation of the thermal performance of a passive solar test-room with wall latent heat storage. Build Environ 32:405–410

Azenha M (2009) Numerical simulation of the structural behavior of concrete since its early ages. School of Engineering University of Porto, Porto

Biswas K, Lu J, Soroushian P, Shrestha S (2014) Combined experimental and numerical evaluation of a prototype nano-PCM enhanced wallboard. Appl.Energy 131:517–529

Building Materials and Products (2007) Hygrothermal properties—tabulated design values and procedures for determining declared and design thermal values

Entrop AG, Brouwers HJH, Reinders AHME (2011) Experimental research on the use of micro-encapsulated phase change materials to store solar energy in concrete floors and to save energy in Dutch houses. Sol Energy 85:1007–1020

Freitas VP, Sousa A, Silva J (2003) Manual de Aplicação de Revestimentos Cerâmicos. APICER—Associação Portuguesa da Indústria de Cerâmica, Coimbra, Portugal

Cunha S, Lima M, Aguiar JB (2016) Influence of adding phase change materials on the physical and mechanical properties of cement mortars. Constr Build Mater 127:1–10

Halawa E, Bruno F, Saman W (2005) Numerical analysis of a PCM thermal storage system with varying wall temperature. Energy Convers Manage 46:2592–2604

Hasan A, Sarwar J, Alnoman H, Abdelbaqi S (2017) Yearly energy performance of a photovoltaic-phase change material (PV-PCM) system in hot climate. Solar Energy 146:417–429

Kenisarin M, Mahkamov K (2007) Solar energy storage using phase change materials. Renew Sustain Energy Rev 11:1913–1965

Kenisarin M, Mahkamov K (2016) Passive thermal control in residential buildings using phase change materials. Renew Sustain Energy Rev 55:371–398

Kong X, Lu S, Huang J, Cai Z, Wei S (2013) Experimental research on the use of phase change materials in perforated brick rooms for cooling storage. Energy Build 62:597–604

Kong X, Yao C, Jie P, LiuY, Qi C, Rong X (2017) Development and thermal performance of an expanded perlite-based phase change material wallboard for passive cooling in building. Energy and Build 152:547–557

Kuznik F, Virgone J (2009) Experimental investigation of wallboard containing phase change material: data for validation of numerical modeling. Energy Build 41:561–570

Lane GA (1983) Solar heat storage: latent heat materials, vol V.1. CRC Press, Boca Raton

Li M, Wu Z, Tan J (2013) Heat storage properties of the cement mortar incorporated with composite phase change material. Appl Energy 103:393–399

Lee KO, Medina MA, Raith E, Sun X (2015) Assessing the integration of a thin phase change material (PCM) layer in a residential building wall for heat transfer reduction and management. Appl. Energy 137:699–706

Lu S, Tong H, Pang B (2018) Study on the coupling heating system of floor radiation and sunspace based on energy storage technology. Energy Build 159:441–453

Manie J (ed) (2010) DIANA TNO User's Manual—Release 9.4.3

Ozisik MN (1993) Heat conduction, 2nd ed

Piselli C, Castaldo VL, Pisello AL (2019) How to enhance thermal energy storage effect of PCM in roofs with varying solar reflectance: experimental and numerical assessment of a new roof system for passive cooling in different climate conditions. Solar Energy 192:106–119

Regulamento das Características de Comportamento Térmico dos Edifícios (RCCTE), D.L. n.º 80/2006 (2006)

Sá AV, Azenha M, de Sousa H, Samagaio A (2012) Thermal enhancement of plastering mortars with phase change materials: experimental and numerical approach. Energy Build 49:16–27

Shamsundar N, Sparrow BM (1975) Analysis of multidimensional conduction phase change via the enthalpy model. J Heat Transfer 97:333–340

Soares N, Costa JJ, Gaspar AR, Santos P (2013) Review of passive PCM latent heat thermal energy storage systems towards buildings' energy efficiency. Energy Build 59:82–103

Zhu N, Wu M, Hu P, Xu L, Lei F, Li S (2018) Performance study on different location of double layers SSPCM wallboard in office building. Energy Build 158:23–31

The manufacturer's authorised representative in the EU is Springer Nature Customer Service Centre GmbH, Europaplatz 3, 69115 Heidelberg, Germany. If you have any concerns regarding our products, please contact ProductSafety@springernature.com

Printed and bound by CPI Group (UK) Ltd, Croydon, CR0 4YY
30/09/2026
02225665-0006